Math

TEACHER'S RESOURCE GUIDE

Level I

ETA Cuisenaire

Vernon Hills, Illinois

VersaTiles® Level 1 Math Teacher's Resource Guide

ISBN 978-1-57162-443-7
ETA 914001R

ETA/Cuisenaire • Vernon Hills, Illinois 60061-1862
800-445-5985 • www.etacuisenaire.com

Printed in the United States of America.
11 12 13 14 15 16 17 18   17 16 15 14 13 12 11 10

Contents

Welcome to VersaTiles®!

Use concrete models!

Numbers and Number Concepts Mascot (red)

VersaTiles® is the unique, individualized mathematics, science, and reading/language arts program that gives your students the freedom to learn, practice, and review concepts at their own pace by matching patterns. Students track their own progress, so they take ownership of their personal learning experience and gain self-confidence as they go.

Think of real-life examples!

Geometry and Measurement Mascot (violet)

VersaTiles® Is a Color-Coded System!

There are eight different titles in this level. Each title focuses on a specific strand. The strands are color-coded so that they can be identified quickly. Here is an overview of the books in this level.

Student Book Number	Book Title	Strand Name	Strand Color
	Level 1 Math Books (Light Blue Level*)		
1	Counting 1–100	Number and Number Concepts	Red
2	Let's ESTIMATE	Estimation and Computation	Green
3	Let's ADD	Estimation and Computation	Green
4	Let's SUBTRACT	Estimation and Computation	Green
5	All About PATTERNS	Patterns, Functions, and Algebra	Blue
6	All About SHAPES and MEASURES	Geometry and Measurement	Violet
7	TIME and MONEY	Geometry and Measurement	Violet
8	TALLIES and TABLES	Statistics and Probability	Orange

Check your answers by estimating!

Estimation and Computation Mascot (green)

* Horizontal stripe in book's upper left corner
Vertical stripe down book's spine

For each strand, there is a mascot who provides strategic hints to help students get started with the activities.

Always look for a pattern!

Patterns, Functions, and Algebra Mascot (blue)

Draw a picture or make a chart!

Statistics and Probability Mascot (orange)

VersaTiles® Jump Start!

Here's one way to get your students started with a *VersaTiles* Lab!

1. Set up the VersaTiles Lab in a five-student Activity Center in your classroom.

2. Divide the class into groups of five so that each group can rotate into the VersaTiles Activity Center.

3. Give each student in the Activity Center a VersaTiles Answer Case and a photocopy of the directions for how to use VersaTiles (page 59).

4. Review the directions and make sure each student understands the concept of VersaTiles.

5. Next, give each student a copy of the Book 1 Test (Formal Assessment, page 75) or the Book 1 Benchmark Activity (Informal Assessment, page 66). You may choose to use both assessment formats or only one. If you use both formats, it is better to administer them one at a time. See pages 25–30 for details about assessment.

6. Decide if you want all students to begin on page 1 of Book 1 or if you want to place students in Book 1 based on assessment results. If you decide to place students individually within Book 1 based on assessment results, then use the Book 1 portion of the Correlation of Tests to Student Activity Books chart on page 83 to determine which activities you want to assign to each student.

7. Give each student a copy of Book 1 and a photocopy of the VersaTiles Student Record Chart (page 85).

8. Have each student mark his/her assigned activities on the Student Record Chart. Then have him/her tape the Student Record Chart to the bottom of the Answer Case (if each student has one) or inside a personal folder to keep track of completed activities.

9. Have a student begin doing the VersaTiles activities independently!

10. When a student has completed Book 1, have him/her hand in the Student Record Chart.

11. Review the Student Record Chart with each student to ensure that he/she has completed the appropriate activities; then encourage the student to place the Student Record Chart in his/her portfolio.

12. Photocopy and administer the VersaTiles Student Self-Assessment Sheet (page 87).

13. Collect and review the completed Self-Assessment Sheets. Then return them to students and encourage students to place them in their portfolios. This would also be a good time to conduct student interviews. (See page 63 for interview ideas.) Or you may wish to administer the formal test again (as a posttest).

14. If students require additional practice with specific skills, you could have them create their own VersaTiles activities that focus on those skills. (See pages 31–32 for ideas.) Then have them challenge others in their group to complete those activities as well.

15. Have students move on to Book 2, and repeat this process beginning at Step 5 above.

Program Overview

VersaTiles® Math is a highly successful, classroom-tested supplemental mathematics program that will build your students' math skills from basic arithmetic through algebra. Students find VersaTiles activities are fun and provide purposeful practice that increases their self-confidence. Teachers appreciate that VersaTiles is self-correcting, non-consumable, and easy to integrate into any mathematics curriculum.

As a teacher, you are a master manager of time and students—yet each of your students learns at his/her own pace. They all need to practice the basics, but each one needs practice in different areas. You need an engaging program that will help you provide meaningful practice for your students and free up your time so that you can focus on their individual needs. VersaTiles is the answer to your classroom needs!

> The ultimate goal of the educational system is to shift to the individual the burden of pursuing his education.
>
> —John W. Gardner

VersaTiles provides a comprehensive list of objectives correlated to specific grade levels with multiple entry points. You can identify the specific practice that individual students need and then direct each student to the appropriate VersaTiles Student Activity Book. When they open the books, students will discover the fun way to learn, practice, and review skills. Students love VersaTiles because the format is engaging and gives them an opportunity to practice the basics as they solve pattern puzzles.

> Children forget what they are taught. They remember what they do.
>
> —Anonymous

As a professional, you know that purposeful practice builds strong mathematical skills. VersaTiles is practice in an enjoyable format that will reinforce basic skills, raise test scores, and build your students' self-confidence.

Grade Level Appropriateness

Each lab focuses on skills that are typically taught at a specific grade level, but keep in mind that each student learns at his/her own pace. So, for example, if you are a fifth grade teacher, you might want to use Lab Levels 4, 5, and 6 and then place each student within the lab and book that is appropriate for him/her based on the results of the Benchmark Activities and/or Formal Tests.

Also remember that there is a spiraling of basic skills from grade level to grade level. However, as the lab number increases, the difficulty level of the skills practiced increases as well. So if your students are functioning at different grade levels, you might want to use more than one lab level and let students work at their own pace to meet their individual needs as they progress through the labs.

Self-Paced Learning

After a brief introduction to VersaTiles, students manage their own learning. They work independently or in small groups, check their own work, and track their own progress (using the Student Record Chart on page 85). This feature of VersaTiles helps students develop a feeling of accomplishment and ultimately helps build students' self-confidence.

With VersaTiles, students do not compete with one another. Each student simply progresses at his/her own individual pace—efficiently, enthusiastically, and confidently.

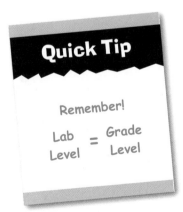

Quick Tip

Remember!

Lab Level = Grade Level

Options for Using VersaTiles®

1. As *Daily Practice* to supplement your math textbook or curriculum

2. As an *Activity Center* to individualize instruction for all students

3. As a *Resource Center* for students who need individualized help with specific skills

4. As *Extra Credit*

5. As *Enrichment* for students who need to be challenged

6. As a *Study Hall Resource*

7. As a *Summer School Program Supplement*

8. As an *ESL Program Supplement*

9. As a *Library Resource*

10. As an *Adult Education Program Supplement*

11. As a *Home Lending Library*

See pages 15–21 for details.

Scope and Sequence Overview

A detailed Scope and Sequence is provided on pages 33–58. Each Student Activity Book in a lab focuses on one of the following five strands.

- **Number and Number Concepts**
- **Estimation and Computation**
- **Patterns, Functions, and Algebra**
- **Geometry and Measurement**
- **Statistics and Probability**

The Student Activity Books are color-coded according to the strands.

Here's an overview of the VersaTiles® Math Program for Grades 1–8.

	Books							
Level	**1**	**2**	**3**	**4**	**5**	**6**	**7**	**8**
1	Number & Number Concepts	Estimation & Computation	Estimation & Computation	Estimation & Computation	Patterns, Functions, & Algebra	Geometry & Measurement	Geometry & Measurement	Statistics & Probability
2								
3					Estimation & Computation	Patterns, Functions, & Algebra		
4								
5								
6								
7								
8								

NCTM Standards Correlation

VersaTiles® correlates directly to the National Council of Teachers of Mathematics (NCTM) Curriculum and Evaluation Standards. NCTM Standards 1 through 4 are integrated throughout all VersaTiles Student Activity Books. NCTM Standards 5 through 13 are addressed in specific VersaTiles Student Activity Books.

The chart below summarizes the correlation between VersaTiles and the NCTM Curriculum and Evaluation Standards.

		VersaTiles Level 1 Books							
		1	2	3	4	5	6	7	8
NCTM Standards for Grades K-4									
1	Mathematics as Problem Solving	★	★	★	★	★	★	★	★
2	Mathematics as Communication	★	★	★	★	★	★	★	★
3	Mathematics as Reasoning	★	★	★	★	★	★	★	★
4	Mathematics as Connections	★	★	★	★	★	★	★	★
5	Estimation	■	■	■	■				
6	Number Sense and Numeration	■	■	■	■	▲		▲	
7	Concepts of Whole Number Operations	■	■	■	■	▲		▲	▲
8	Whole Number Computation	■	■	■	■	▲			▲
9	Geometry and Spatial Sense		▲			■	■		▲
10	Measurement		▲				■	■	▲
11	Statistics and Probability								■
12	Fractions and Decimals							■	
13	Patterns and Relationships	▲	▲	▲	▲	■	▲	▲	▲

★ Topic is integrated throughout all the books.   ■ Topic is the main topic of the book.   ▲ Topic also appears in the book.

The NCTM Curriculum and Evaluation Standards recommend that all students have access to calculators and computers. However, this does not mean that students do not need to achieve proficiency with paper-and-pencil computation. These mechanical tools were created to expedite the process of calculation, not replace the skills necessary to understand how and why calculation is done. In fact, calculators and computers have so dramatically increased the speed at which basic computation can be done, it is essential that students today have number sense and the ability to estimate and compute to determine whether or not an answer is reasonable. Students need to learn how to choose the appropriate tools for a variety of different problem-solving situations, and that's what VersaTiles is all about!

The Benefits of VersaTiles®

Benefits for Students

❧ **VersaTiles is engaging.** VersaTiles captures students' attention, and they like it!

❧ **VersaTiles is a fun way to practice.** Students have the opportunity to practice skills and concepts encountered on standardized tests in a fun context.

❧ **VersaTiles is self-correcting.** Students get immediate feedback on their work, so they know what they got right and can correct what they didn't.

❧ **VersaTiles helps build self-confidence.** It is used independently, so students can work at their own pace and they don't need to compete with anyone but themselves.

❧ **VersaTiles matches your mathematics curriculum.** VersaTiles addresses everything from basic arithmetic to spatial relationships to algebraic concepts to probability and statistics, so students can get the kind of practice they need!

Tell me and I'll forget. Show me, and I may not remember. Involve me, and I'll understand.

—*Native American Saying*

Estimation and Computation Mascot

Benefits for Teachers

❚ **VersaTiles® is correlated to the NCTM Standards and to current textbook programs.** Students are getting purposeful practice that is geared toward the standard curriculum for their grade level. VersaTiles is designed to provide reinforcement for skills in the sequence in which they are taught in most basal programs and as recommended by the National Council of Teachers of Mathematics (NCTM).

❚ **Objectives are listed at the bottom of each activity page.** You know the exact skill level of each activity, so you can maximize practice time for students!

❚ **There are many ways to use and extend VersaTiles.** VersaTiles is meaningful for all students, and it fits into any classroom setting.

❚ **Students can work independently with VersaTiles.** While students work independently, you are free to focus on important tasks such as performance-based assessment and individualized instruction.

❚ **VersaTiles addresses the full range of mathematics skills for grades 1–8.** With VersaTiles, students can get the practice they need, whenever they need it. Use VersaTiles for remediation, practice, and enrichment, too!

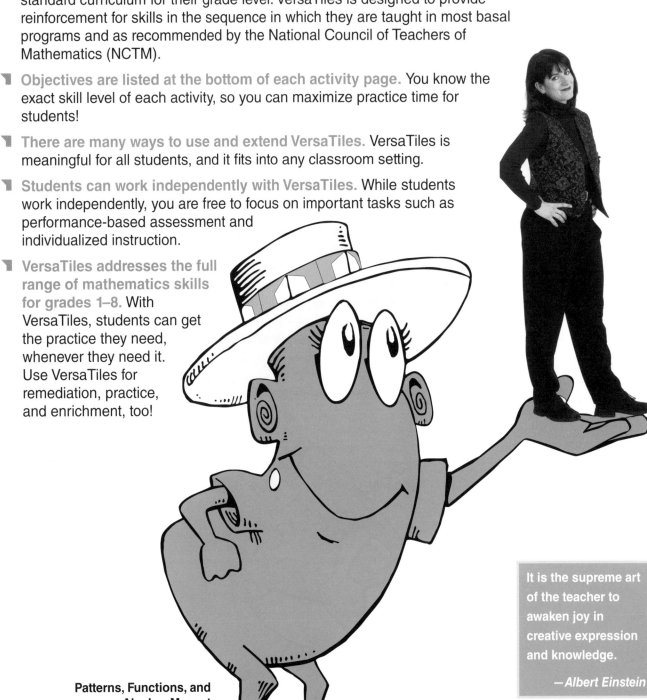

Patterns, Functions, and Algebra Mascot

It is the supreme art of the teacher to awaken joy in creative expression and knowledge.

—*Albert Einstein*

What's in a VersaTiles® Math Lab?

Each lab contains—

40 Student Activity Books
Each lab contains 5 copies each
of 8 Student Activity Books.

1 Teacher's Resource Guide This book provides an overview of the VersaTiles program along with a wealth of ideas for assessing students and implementing the program. It includes a complete scope and sequence for the program, both informal and formal assessment options, and abundant blackline masters so that you can customize VersaTiles to your teaching style and individual needs.

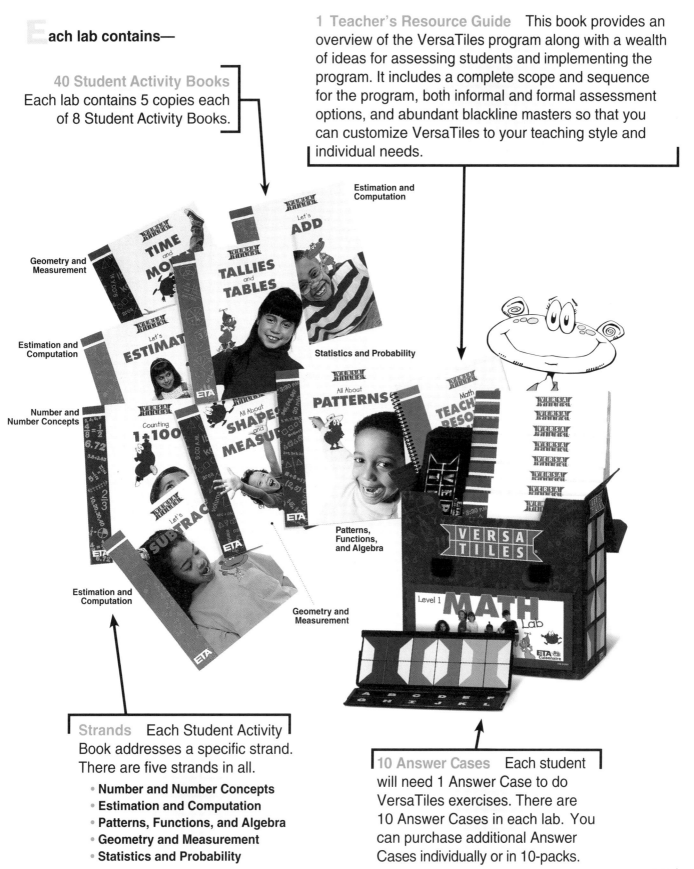

Estimation and Computation

Geometry and Measurement

Estimation and Computation

Number and Number Concepts

Statistics and Probability

Patterns, Functions, and Algebra

Estimation and Computation

Geometry and Measurement

Strands Each Student Activity Book addresses a specific strand. There are five strands in all.

- **Number and Number Concepts**
- **Estimation and Computation**
- **Patterns, Functions, and Algebra**
- **Geometry and Measurement**
- **Statistics and Probability**

10 Answer Cases Each student will need 1 Answer Case to do VersaTiles exercises. There are 10 Answer Cases in each lab. You can purchase additional Answer Cases individually or in 10-packs.

What's in the Student Activity Books?

Exercises   Each Activity consists of 12 exercises. The answer to each exercise can be found in one of the cells in the Answer Box at the bottom of the page. The student reads the exercise and then looks for the correct answer in the Answer Box. When the student finds the correct answer, he/she places the tile in the Answer Case, matching the exercise number to the letter of the cell in the Answer Box that contains the correct answer.

Mascots   provide strategic hints to help students get started!

Objectives   at the bottom of each page state the purpose of the activity and the specific skill that is being practiced. This will help you ensure that all students are engaged in purposeful practice. You can also use these objectives to create your own VersaTiles® activities.

Use the scope and sequence on pages 33–58 to locate specific topics within the program.

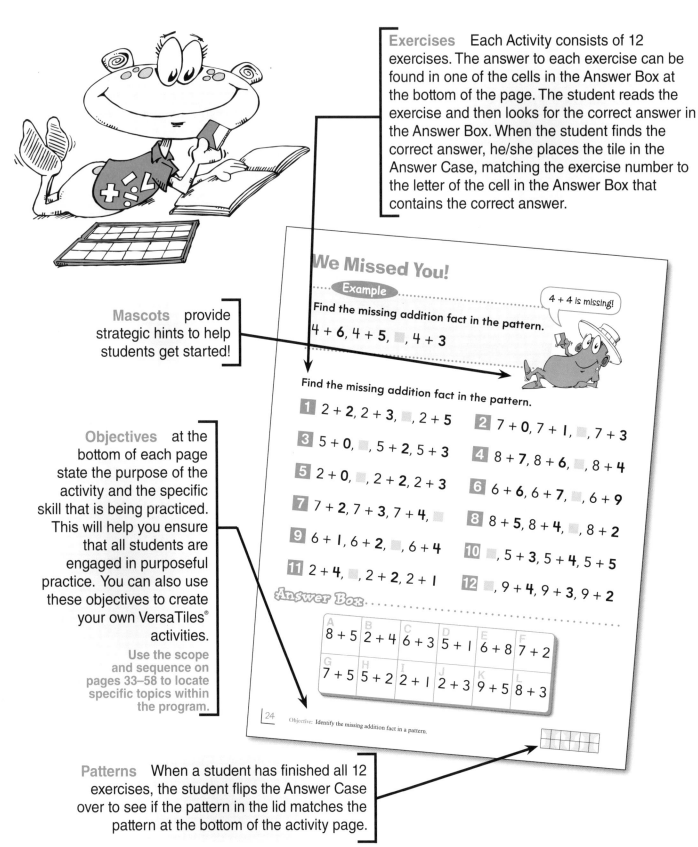

We Missed You!

Example

Find the missing addition fact in the pattern.

4 + **6**, 4 + **5**, ■, 4 + **3**

4 + 4 is missing!

Find the missing addition fact in the pattern.

1 2 + **2**, 2 + **3**, ■, 2 + **5**

2 7 + **0**, 7 + **1**, ■, 7 + **3**

3 5 + **0**, ■, 5 + **2**, 5 + **3**

4 8 + **7**, 8 + **6**, ■, 8 + **4**

5 2 + **0**, ■, 2 + **2**, 2 + **3**

6 6 + **6**, 6 + **7**, ■, 6 + **9**

7 7 + **2**, 7 + **3**, 7 + **4**, ■

8 8 + **5**, 8 + **4**, ■, 8 + **2**

9 6 + **1**, 6 + **2**, ■, 6 + **4**

10 ■, 5 + **3**, 5 + **4**, 5 + **5**

11 2 + **4**, ■, 2 + **2**, 2 + **1**

12 ■, 9 + **4**, 9 + **3**, 9 + **2**

Answer Box

A	B	C	D	E	F
8 + 5	2 + 4	6 + 3	5 + 1	6 + 8	7 + 2
G	H	I	J	K	L
7 + 5	5 + 2	2 + 1	2 + 3	9 + 5	8 + 3

24

Objective: Identify the missing addition fact in a pattern.

Patterns   When a student has finished all 12 exercises, the student flips the Answer Case over to see if the pattern in the lid matches the pattern at the bottom of the activity page.

How Can I Use VersaTiles® in My Classroom and School?

One of the wonderful features of VersaTiles is that it is, in fact, versatile! Here are some of the ways you can use it.

VersaTiles® as Daily Practice

If you want your students to get skills practice on a daily basis without the boredom of "drill and kill," then you will want to have a VersaTiles Answer Case for each student along with a variety of Student Activity Books. This gives you the opportunity to have "VersaTiles Time" on a daily or regularly scheduled basis.

You will find that the program is closely tied to your curriculum and addresses the skills tested in standardized tests, so you know that your students are getting quality practice time with VersaTiles.

There are many ways to use the VersaTiles program in your class. You need to decide which labs best suit your students, then decide how many labs you need. For example, if you are a first grade teacher with a diverse class in which some students are functioning at grade level while others are above or below grade level, you might want to use Lab Levels 1 and 2. This gives you a total of 16 different book titles (8 titles at each level) and 20 Answer Cases.

You might want to sort the class into groups according to the grade levels at which they are currently performing. There are five copies of each title in each lab, so five students can use the same book simultaneously.

On the other hand, if most students are functioning at the same grade level, you might want to provide only Lab Level 1. If you want all students to be able to use the lab at the same time, you will probably need to have two or more Lab Level 1 sets, depending on your class size.

> From the very beginning of his education, the child should experience the joy of discovery.
>
> —*Alfred North Whitehead*

VersaTiles® as Activity Centers

VersaTiles is ideal for Activity Centers. You can set up a corner in your classroom where your students will have easy access to the VersaTiles Lab(s). Have groups of five (or fewer) students rotate into the Activity Center at various times throughout the day (or week).

Encourage students to use the Activity Center at assigned times, when they have completed class assignments, or when you have scheduled practice of a specific skill.

The Activity Centers can also be used to help you place students in the appropriate Student Activity Books so that practice is purposeful. For example, to use VersaTiles for informal assessment purposes, assign the class individual quiet work and have groups of three to five students take turns rotating into the Activity Center. Observe students in the Activity Center as they complete a specific benchmark activity (see pages 66–73). You can use the Observation and Interview Notes on page 64 to record your observations.

VersaTiles® in Resource Centers

If you have a Resource Center in your school, VersaTiles belongs there too! A Resource Center is typically a place where students go to receive individualized instruction from a Special Education Teacher or a tutor. Students usually visit a Resource Center when they are having difficulty with a particular subject or concept, or when they are classified as students with special needs. VersaTiles can be an effective management tool in Resource Centers. For example, assuming a student visits the Resource Center for a full class period, he or she may spend a good portion of that time working one-on-one with a teacher or tutor. VersaTiles fits into the Resource Center as a follow-up to the one-on-one time. After teacher and student practice a specific skill or concept together, the teacher can then use the VersaTiles Scope and Sequence (see pages 33–58) to assign a VersaTiles activity or activities that reinforce the newly acquired skill. Because VersaTiles makes practice fun and captures the student's attention, it is ideal for students with special needs, giving them a feeling of independence and satisfaction. The self-correcting feature of VersaTiles is especially good for students who are struggling with basic math skills because it helps them develop self-confidence as mathematical problem solvers.

Estimation and Computation Mascot

How Can I Use VersaTiles® in My Classroom and School?

1. As *Daily Practice* to supplement your textbook program or curriculum

2. As an *Activity Center* to individualize instruction for all students

3. As a *Resource Center* for students who need individualized help with specific skills

4. As *Extra Credit*

5. As *Enrichment* for students who need to be challenged

6. As a *Study Hall Resource*

7. As a *Summer School Program Supplement*

8. As an *ESL Program Supplement*

9. As a *Library Resource* to enhance the school library offerings

10. As an *Adult Education Program Supplement*

11. As a *Home Lending Library* to communicate with parents about students' learning experiences.

VersaTiles® as Extra Credit

Encourage students to do VersaTiles for extra credit! Simply photocopy and use the VersaTiles Student Record Chart (page 85) to keep track of student progress. You might want to have students create a VersaTiles Extra Credit Section in their portfolios so that they can keep track of their progress. VersaTiles for extra credit is an excellent way to help students improve their grades with meaningful, yet fun, practice. You might also encourage students to write their own VersaTiles Activities for extra credit. This can be challenging and very rewarding. Refer to pages 31–32 for details.

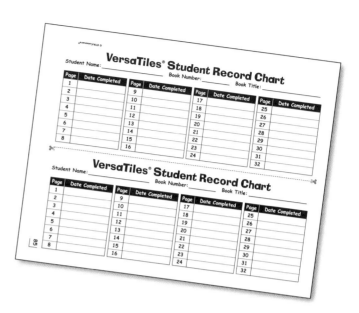

VersaTiles® as Enrichment

If, like most classes, your class is heterogeneous, then you probably find that some students in your class are finished with their work before the other students. Even these students need practice, and like all students, they are motivated by the feeling of accomplishment that comes with completing a finite task! VersaTiles provides you with a tool that will motivate and challenge them. Further, if you set up VersaTiles in your classroom as something students are entitled to do after they finish their regular class work, then you will find that all students will strive to complete their class work more quickly so that they can do VersaTiles. VersaTiles is gamelike, so students have fun as they learn!

VersaTiles® as a Study Hall Resource

How many times have you had students walk into study hall with nothing to do? How many times have you had students tell you they are bored and need something to do during free time? VersaTiles is the answer. Simply set up a lab at the back of the class and let the students decide when they want to use it. You won't be disappointed with the results! Make sure students return all materials at the end of each Study Hall period. Photocopy the Student Record Chart (page 85) and have students use it to keep track of their own progress. You might want to keep the Student Record Charts in a file so that the records are accessible to students each time they enter Study Hall.

> If a man does not keep pace with his companions, perhaps it is because he hears a different drummer. Let him step to the music which he hears, however measured or far away.
>
> —Henry David Thoreau

VersaTiles® as a Summer School Program Supplement

Every teacher who has ever done it knows that summer school is the toughest kind of school to teach. It's usually hot outside, and students get edgy at the thought of going to school in the summertime to study a subject that has probably meant "failure" to them in the past. But VersaTiles can lighten the load and provide a valuable summer school experience for both students and teachers. The fun format and direct correlation of VersaTiles activities to basic skills makes VersaTiles and summer school a winning combination.

VersaTiles® as an ESL Program Supplement

VersaTiles can be a fun way for students for whom English is a second language to practice basic reading and math skills. VersaTiles is perfect for ESL students because the VersaTiles "multiple-choice" format gives students with limited English vocabularies a finite number of answer possibilities. Providing a choice of answers not only reinforces vocabulary, but also enhances problem-solving skills, as the student is free to focus on finding the solution.

> If we succeed in giving the love of learning, the learning itself is sure to follow.
>
> —John Lubbock

VersaTiles® in the School Library

Many schools have purchased VersaTiles as an additional resource for the school library because students enjoy it and ask for it so often. We recommend this as an extension to the classroom use of VersaTiles.

VersaTiles® in Adult Education Programs

VersaTiles is also used in Adult Basic Education (ABE) Programs. The VersaTiles Math Program addresses everything from simple addition, subtraction, multiplication, and division to pre-algebra and basic geometry.

Statistics and Probability Mascot

Creating a Home Lending Library with VersaTiles®

When parents do VersaTiles activities with their children, they get first-hand information about their child's knowledge base. It's also an easy way for you to show parents that their children are learning basic skills as well as problem solving in math class.

Here's how you can set up a VersaTiles Home Lending Library in your classroom.

Purchase a VersaTiles Lab. Each lab has 40 Student Activity Books so you will also need 40 large size, plastic zip-top-type freezer bags and 40 Answer Cases. Place 1 Answer Case, 1 Student Activity Book, and 1 copy of the Home Lending Library Tracking Form #1 (page 60) in each bag. You will have 40 packets in all.

Obtain blank, 3.5" x 5", white, stick-on labels to label each bag. Number the bags consecutively starting with the number 1. On each label, list the packet number and the contents of the packet.

You can have students help you create these packets at the beginning of the school year. Then make a copy of the Home Lending Library Wall Chart (page 62) and hang it on the wall in your classroom. You might want to enlarge the form on a photocopying machine before you hang it up. Each time you send VersaTiles home with your students, have a responsible student complete the chart as you hand out individual packets to other students in the class.

Home Lending Library Tracking Form #1 (page 60) is a means by which parents can communicate with you about their child's learning experiences at home with VersaTiles. Consider leaving this form in the packet as it travels to students' homes. It promotes parental involvement and gives parents an opportunity to see what other parents have had to say.

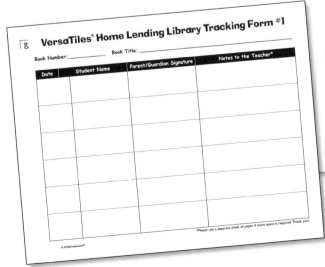

Quick Tip

Before using the Home Lending Library Wall Chart, you might want to fill in all the packet numbers first, in numerical order, so that it is easy to see at a glance which packets are in and which packets are out. Use the Notes section on the Wall Chart to make remarks about the condition and contents of the packets as they are checked in and out.

Give each student a copy of Home Lending Library Tracking Form #2 (page 61) at the beginning of the school year and direct him/her to keep it in a safe place for repeated use. Each time a student takes a packet home, he/she should record the packet number and date on the form. Then the student should circle the pages corresponding to the activities completed. When finished, the student should return the packet and record the return date on the form. Be sure to supervise the return procedure to ensure that all materials have been properly returned. It's a good idea to use the Home Lending Library Wall Chart (page 62) in conjunction with this form so that you have accurate records of where the materials are.

VersaTiles® Home Lending Library Tracking Form #2

Student Name: _____

Book Number	Book Title	Packet Number	Date Out	Date In	Circle the pages you completed.
1	Counting 1-100				1 2 3 4 5 6 7 8 9 10 11 12 13 14 15 16 17 18 19 20 21 22 23 24 25 26 27 28 29 30 31 32
2	Let's ESTIMATE				1 2 3 4 5 6 7 8 9 10 11 12 13 14 15 16 17 18 19 20 21 22 23 24 25 26 27 28 29 30 31 32
3	Let's ADD				1 2 3 4 5 6 7 8 9 10 11 12 13 14 15 16 17 18 19 20 21 22 23 24 25 26 27 28 29 30 31 32
4	Let's SUBTRACT				1 2 3 4 5 6 7 8 9 10 11 12 13 14 15 16 17 18 19 20 21 22 23 24 25 26 27 28 29 30 31 32
5	All About PATTERNS				1 2 3 4 5 6 7 8 9 10 11 12 13 14 15 16 17 18 19 20 21 22 23 24 25 26 27 28 29 30 31 32
6	All About SHAPES and MEASURES				1 2 3 4 5 6 7 8 9 10 11 12 13 14 15 16 17 18 19 20 21 22 23 24 25 26 27 28 29 30 31 32
7	TIME and MONEY				1 2 3 4 5 6 7 8 9 10 11 12 13 14 15 16 17 18 19 20 21 22 23 24 25 26 27 28 29 30 31 32
8	TALLIES and TABLES				1 2 3 4 5 6 7 8 9 10 11 12 13 14 15 16 17 18 19 20 21 22 23 24 25 26 27 28 29 30 31 32

61

Quick Tip

Before you send any VersaTiles® materials home with students, you will probably want to send parents a brief letter to introduce them to the concept of the Home Lending Library and to get them to help you manage the flow of materials.

Dear Parent or Guardian:

We will be using a new, innovative program called *VersaTiles* in class this school year. *VersaTiles* provides your child with opportunities to practice basic skills and problem solving in a fun way! We will use *VersaTiles* in class to supplement our textbook program. Your child will also bring *VersaTiles* home periodically.

Please ask your child to show you how *VersaTiles* works. Then take a few minutes to try an activity or two with your child. If you wish to share your experience with me, write me a note in the "notes" section on Home Lending Library Tracking Form #1.

If you have any questions, please call me at _____. It is probably easiest to reach me between the hours of ____ and ____. If you have difficulty reaching me, please leave a detailed message on my voicemail with your phone number and the best time to call you.

I look forward to working with you to provide your child with an exceptional learning experience this school year.

Sincerely,

_____
Teacher

Involving your students in the organization and management of a Home Lending Library can be very rewarding. For example, when students return the packets, one student can check that all the appropriate contents of the packet are in the bag, while another student completes the Home Lending Library Wall Chart (page 62). You will want to observe as these transactions take place.

When students help you manage the Home Lending Library, it can save you time and frustration and teach your students important lessons about organization and teamwork.

Quick Tip

You might want to store your Home Lending Library Packets in a mailbox storage system or some type of organized file system in which you can quickly see which packets are "in" and which packets are "out." The biggest challenge in managing a Home Lending Library is organization, so you need to develop a strict system for tracking the materials.

How Can I Create a Home Lending Library?

To start your VersaTiles Home Lending Library, first decide how you want to organize and use it. For example, you may want to send packets home with only five (or fewer) students at a time so that it is easier to keep track of the packets. Group A might take the packets home on Monday, while Group B takes the packets home on Tuesday, and so on. This way each group can have the opportunity to take home the same packet within the same week. Since the books correlate to your curriculum, this organization of the Home Lending Library can correspond with your lesson plans.

VersaTiles Starter Sets for Home Use

VersaTiles Starter Sets are ideal for home use. You might encourage your Parent Teacher Association (PTA) to purchase VersaTiles Starter Sets for this purpose. A VersaTiles Starter Set contains one copy of each Student Activity Book in the Lab and one Answer Case, so you can send the VersaTiles Starter Sets home and have students keep them there. This minimizes the wear and tear on the packets caused by the back and forth travel between school and home.

Classroom Management Tips

Individualizing Instruction

Tracking each student's progress will provide you with valuable information about individual skill development. In order to do this, have students complete the VersaTiles Student Record Chart (page 85) as they go. You may want to photocopy this chart and have students tape it to the back of their Answer Cases before they begin each new Student Activity Book.

Before beginning a new book, assess each student to decide which pages within the book provide the most appropriate level of practice for that student. Circle those activity pages on the Student Record Chart, or have the student do so. Then, as the student completes the activities, have the student write the date next to the page number to record his/her progress on the assigned activities.

When a student has completed all the assigned activities within a specific book, have the student share the completed Student Record Chart with you. You might also administer the VersaTiles Student Self-Assessment Sheet (page 87) upon the completion of each book. It provides insight into students' dispositions and can be a nice addition to students' portfolios. This might be a good time to interview the student to determine what the student has learned from completing these activities and to find out whether or not the student requires additional instruction or practice. Or you might want to administer the Formal Test for the book (as a post-test) to verify whether the student has mastered the skills in the book.

A VersaTiles Class Record Chart is provided on page 86. You can use this chart to keep track of the entire class. Just make one copy of this chart and write each student's name on it. Then make eight copies of the chart containing all students names—one chart for each book in the lab. As students progress through the program, you can track their progress. This will tell you at a glance which students have completed which activities. This chart is particularly helpful if you have several students working in different books simultaneously.

Flexible Grouping

VersaTiles® activities are usually used independently due to their self-correcting format, but the activities can be equally effective when used in other ways. For example, students can work in pairs or small groups on VersaTiles, helping one another out. This is especially true of the problem-solving activities which are integrated throughout the VersaTiles books and which require reading comprehension skills.

There are several grouping options. For example, you can divide the class into teams and host a VersaTiles competition where each team completes the exercises as a joint venture. Give each team points based on how many activities they complete correctly within a given period of time. Tally the points and figure out who the "champs" are at the end of the week. To make team competitions more interesting and challenging, use activities that do not contain patterns at the bottom of the page, so that students cannot check their own work. It will also make it easier for you to determine which team is the winner! You can use the Benchmark Activities (pages 66–73) for this purpose, or you can create your own VersaTiles activities. Be sure to have the teams record their answers using photocopies of the VersaTiles Work Slate (page 89), so that you have a written record of their solutions.

The Ultimate VersaTiles® Classroom

If you want all of the students in your class to work on VersaTiles simultaneously and to move through the VersaTiles books consecutively, then you will need to provide several labs sets for the class. For example, if you have 25 students in your class, you will need 5 lab sets in order for all students to use the same book at the same time. This will provide you with 50 Answer Cases (since there are 10 Answer Cases in each lab). Consider giving each student two cases—one for school use and one for home use (another variation of the Home Lending Library concept). This is not uncommon, but it does require a substantial budget. Some schools have obtained help from their PTA (Parent Teacher Association) in implementing a program such as this. The PTA recognizes the value of VersaTiles because they immediately see that it is fun and gives parent and student a tool with which to build skills.

> The pupil who is never required to do what he cannot do, never does what he can do.
>
> —*John Stuart Mill*

Writing in Mathematics

Some of the VersaTiles activities can be completed mentally without using paper and pencil (for example, estimation and computation activities), but many other VersaTiles activities require students to draw pictures, make tables, calculate, or trace through logical steps using pencil and paper. Encourage students to write as they complete the VersaTiles activities.

When students write about mathematical ideas, they clarify their thinking and gain deeper insight into those ideas. Because many students are visual learners, it is best to encourage them to use scratch paper as needed. You may even encourage students to take notes about questions they may have as they complete the activities. Encourage them to write in their journals as they go. For example, they might write about a new math algorithm, procedure, or shortcut they discovered by doing a specific VersaTiles activity.

An optional VersaTiles Work Slate is provided on page 89. You may wish to photocopy this Work Slate and give it to students so that they can write out solutions to the exercises on the Work Slate in addition to actually moving the tiles in the Answer Case. The Work Slate can be especially helpful to students who make careless errors, either because they are hurrying or because they have difficulty making the translation from mental calculations to the physical placement of the tiles. After all, VersaTiles is a multi-task application. Work Slates are also useful for team competitions, as described on page 22.

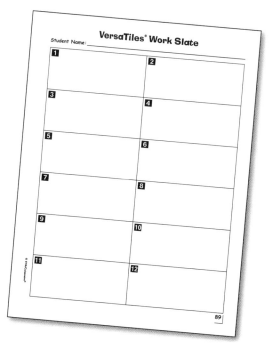

VersaTiles® Helps Improve Standardized Test Scores!

The "multiple-choice" format of VersaTiles resembles the format of standardized tests. We all know that test taking is a learned skill. Experience with the VersaTiles format, helps students become better at taking multiple-choice tests. For example, with VersaTiles, students use the answers provided to help solve the problems, eliminate obvious wrong answers, and estimate to narrow down the possible correct answers.

VersaTiles also helps students develop a system for keeping track of their responses. When using VersaTiles, students complete exercises, search for and find answers, and record responses (by placing each tile in the correct cell in the Answer Case). This gives students an opportunity to practice their eye-hand coordination—a skill that is so important on standardized tests. Facility with this skill can mean the difference between success and failure for even the most mathematically adept students on these types of tests.

Using VersaTiles to Write IEPs (Individualized Education Programs)

Public Law 105–17, Individuals with Disabilities Education Act (IDEA), requires that a written Individualized Education Program (IEP) be completed for each qualified student. An IEP is designed to ensure that, when a child requires special education, the program is appropriate for the child's needs. Once major goals are established, specific objectives must be written and monitored. The specific learning objectives for the VersaTiles program are written in detail so that teachers, aides, and parents can carry out effective instruction.

The VersaTiles Math Program provides a great amount of assistance. Each book is arranged and written with very specific objectives that are in a hierarchical order within each strand. As you progress through the books, you will find that the objectives become more sophisticated. The books cover the main objectives addressed in a typical math curriculum. The Scope and Sequence on pages 33–58 can assist you in writing objectives and assigning activities, and the Student Record Chart (page 85) can serve as a management system (which is required in Public Law 105–17).

Using the Certificate of Completion

A Certificate of Completion is provided on page 88. Use this to motivate students and reinforce the feeling of accomplishment students experience upon completing an assignment. The Student Record Chart (page 85) serves as a learning contract. When the learning contract is fulfilled, the Certificate of Completion is the reward!

Authentic Assessment in the VersaTiles® Program

Authentic Assessment is an important component of the VersaTiles program because it helps to ensure purposeful practice. Authentic Assessment involves examining a student's process as well as his/her work product in order to obtain a holistic perspective of the student's progress. Students' performance can be assessed based on these five categories:

- Mathematical Reasoning
- Mathematical Tools and Techniques
- Mathematical Understanding
- Mathematical Communication
- Mathematical Dispositions

Further, students' performance can be assessed using informal (interviews, observations, self-evaluations) as well as formal (tests and work products) assessment. Options for both informal and formal assessment are provided on the pages that follow. Using a combination of informal and formal assessment enables you to observe students' processes as well as their products.

Assessing Students' Process

Here are some ways to determine how students are working and thinking:

1. Observe students as they complete VersaTiles activities.

2. Ask students questions, either informally while they are working or more formally, using a pre-planned set of questions, as in interviews.

3. Examine samples of students' writing in which they explain the approaches, strategies, and procedures they employed to solve problems or complete VersaTiles activities.

4. Listen to students' verbal explanations of how they solved problems or completed VersaTiles activities.

5. Observe students' affective experiences such as enjoyment, pride, accomplishment, frustration, confusion, and so on.

The following black-line masters are provided to help you assess your students' processes:

1. Observation and Interview Guide (page 63)

2. Observation and Interview Notes (page 64)

3. Benchmark Activities from VersaTiles Math, Books 1–8 (pages 66–73)

Assessing Students' Work Product

Students' work products are the result of assignments. They are the physical examples of students' work. Examples of students' work products in the VersaTiles® program include:

1. Written results of the Formal Tests for VersaTiles Math

2. Journal entries about students' experiences with VersaTiles

3. Reflective writing assignments, such as Student Self-Assessments

4. Scratch paper on which students show their work on specific VersaTiles activities

5. Work Slates on which students write answers to specific VersaTiles activities

6. Models or drawings students create as they think through specific VersaTiles activities

7. Student-created VersaTiles activities

The following blackline masters are provided to help you assess students' work products:

1. Formal Tests for VersaTiles Math (See pages 75–82 for one formal test for each book in the Lab)

2. VersaTiles Student Self-Assessment Sheet (page 87)

3. VersaTiles Work Slate (page 89)

4. Blank VersaTiles Templates (pages 90–91)

> If the school sends out children with a desire for knowledge and some idea of how to acquire and use it, it will have done its work.
>
> —*Sir Richard Livingstone*

Portfolios

A portfolio is an organized collection of student work that can be shared with others. It contains samples of the student's work that are relevant to the student and that demonstrate the student's growth or change over a period of time. An important feature of a portfolio is that it contains work that is selected by the student.

A portfolio may contain samples of a student's work on VersaTiles. For example, a student may include the following VersaTiles work products in his/her portfolio: Student Record Charts, Work Slates, Student Self-Assessment Sheets, Formal Tests, samples of the VersaTiles activities the student has written, scratch paper containing interesting shortcuts or algorithms that he/she discovered while solving specific problems, and samples of journal entries.

Informal Assessment

As you observe your students working every day, you take a pulse on their progress and are keenly aware of each individual's strengths and weaknesses. You know the areas in which your students need practice. To help you verify your observations, we have provided means by which you can evaluate your students informally to ensure that the practice you assign is purposeful.

Benchmark Activities

For each Student Activity Book in the lab, there is a corresponding Benchmark Activity (see pages 66–73) that can help you determine students' knowledge of benchmark skills that are representative of the main skills addressed in that book. The Benchmark Activity is pulled directly from the VersaTiles® Student Activity Book. The only difference is that the Benchmark Activity does not contain a pattern at the bottom of the page, so students cannot check their own work—you have to check it for them. Administer the Benchmark Activity before the student begins working in a specific VersaTiles Student Activity Book, so that you can be confident that the practice you are assigning is purposeful.

Observation and Interview Guide and Notes

You will probably want to administer the Benchmark Activities to individuals or to small groups of students using an observation and interview approach. Before you begin, make one copy of the Observation and Interview Notes (page 64) for yourself. Review the Benchmark Activity (see pages 66–73), and use the Observation and Interview Guide (page 63) to come up with a few key questions for each category. Write the questions in the appropriate spaces on the photocopy of the Observations and Interview Notes. Then make one copy of the appropriate Benchmark Activity and your Observation and Interview Notes for each student you plan to evaluate.

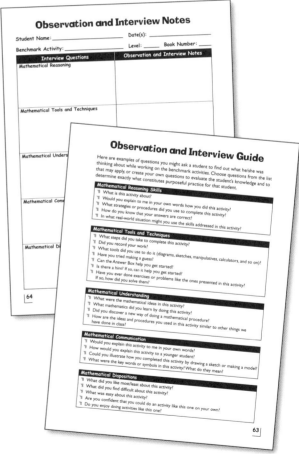

Observe the student's thinking and take notes as the student completes the activity. When the student has completed the activity, use the patterns on page 65 to check his/her work. If the student made any errors, review those exercises with the student and guide him/her to discover the correct answers. Simultaneously, use the interview questions you wrote to informally assess students and record your observations.

After you have completed your evaluation, refer to the Scope and Sequence Chart on pages 33–58 to decide which pages students should complete within the specific VersaTiles Student Activity Book. The Scope and Sequence is organized in clusters so you can see at a glance which skills are covered in each Student Activity Book.

VersaTiles® Student Record Chart

Photocopy the Student Record Chart (page 85) for each student before he/she begins working in a specific VersaTiles Student Activity Book. Assign the pages you want the student to complete, and have the student circle those pages on his/her Student Record Chart. Then you can have the student tape the Student Record Chart to the bottom of his/her Answer Case so it won't get lost. Have the student check off (or mark the date next to) each activity on this form as he/she completes it.

When the student has completed all the assigned activities within a given book, have the student hand in the completed Student Record Chart. Give the student a Certificate of Completion (page 88). This might be a good time to conduct another informal interview or administer another Benchmark Activity. You can make your own Benchmark Activities, too! Simply choose an activity from a Student Activity Book that is significant and appropriate—one that you think is representative of what the student should be able to do. Photocopy it, but hide the pattern when you copy it so the student cannot self-check his/her work. Administer the activity and observe the student as he/she completes it.

Student Self-Assessment

Upon completion of a specific VersaTiles Student Activity Book, you might also have the student complete the VersaTiles Self-Assessment Sheet (page 87). Be sure to collect and review the completed Self-Assessment Sheet, because it will give you insight into the student's disposition as well as the student's knowledge of mathematics. You might even review it with the student or suggest that the student save it in his/her portfolio.

Formal Assessment

Formal Assessment is provided in the VersaTiles® program to help you assign practice that corresponds to the individual needs of each student. When formal assessment is used in conjunction with informal assessment, you can get an accurate picture of your students' knowledge and understanding, which, in turn, will help you facilitate student learning.

Pre-tests and Post-tests

Pages 75–82 of this Teacher's Resource Guide contain a formal, pencil-and-paper test for each book in the VersaTiles Lab. The tests consist of a combination of fill-in-the-blank, multiple-choice, and open-ended questions. The questions match the objectives addressed in the corresponding VersaTiles books. There are several ways to use the tests. For example, you can use the formal tests as pre-tests and/or post-tests in the following way:

1. Before a student begins working within a specific VersaTiles Student Activity Book, administer the formal test that corresponds to that book.

2. Score the test.

3. Use the test results and the Correlation of Tests to Student Activity Books chart on pages 83–84 to determine which activities the student should complete.

4. Return the test to the student and review the results with him/her. Then have the student save the test in a safe place, such as his/her portfolio, for future reference.

5. Assign specific activities to the student based on the test results.

6. When the student finishes the assigned activities, administer the test again.

7. Score the test.

8. Review the results of the test with the student.

9. Have the student compare the results of the post-test with the results of the pre-test.

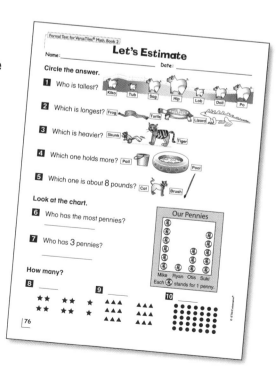

Student-Created VersaTiles® Activities

If a student still needs practice in specific areas after completing a VersaTiles Student Activity Book, consider having the student write his/her own VersaTiles activities to match the objectives with which the student needs practice. Writing a VersaTiles activity requires a clear understanding of the objective and, as a result, provides meaningful practice. Further, students can challenge their classmates to complete the VersaTiles activities they create. Templates are provided on pages 90–91.

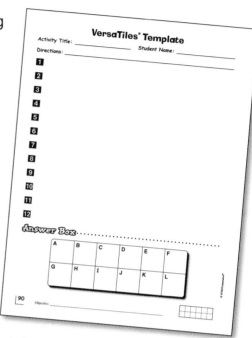

Journal Entries

Encourage your students to write in their journals about their experiences with VersaTiles. Writing in this way promotes critical thought and also provides you with information about students' knowledge and dispositions.

Portfolio Suggestions

Encourage students to select representative samples of their work for their portfolios. Make sure they include samples that represent the full range of their learning experience. For example, they should include work that is excellent, as well as work that needs improvement. For work that needs improvement, they should include first drafts, as well as revised work, to illustrate their progress over time. Most importantly, they should include work that is meaningful to them.

Portfolios are ideal tools for demonstrating students' progress in parent conferences. Because the work samples that students place in their portfolios are self-selected, it is likely that students will have mentioned some of those work samples to their parents throughout the course of the school year. Consequently, parents may be familiar with some of the work you show them. In any case, portfolios reveal important information about students' learning experiences.

> By learning you will teach; by teaching you will learn.
>
> —*Latin Proverb*

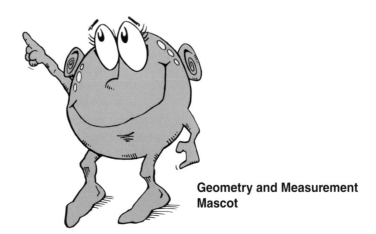

Geometry and Measurement Mascot

Extending VersaTiles®

How to Create Your Own VersaTiles® Activities

Templates are provided on pages 90–91 so that you can create your own VersaTiles activities. It takes some practice to become proficient at writing VersaTiles activities, but after you try it a few times, you will find that it is an easy way to provide your students with motivating practice.

Here's how to write a VersaTiles activity:

- Think of each VersaTiles Activity as one big multiple-choice problem. You need 12 questions and 12 answers. You also need good distractors, just like on any multiple-choice test.

- After you decide on the objective for the activity, write 6 questions. As you write the first 6 questions, record the answers, then record a distractor (or wrong answer) for each question. Use those distractors as the answers to the next 6 questions.

- You might want to adapt some of your favorite practice sets as you begin creating your own VersaTiles activities. Don't worry about putting the answers in order at first. Instead concentrate on writing appropriate exercises. Then, when you are finished writing the exercises, use the patterns and pattern codes provided on pages 92–93 to put the answers in a specific order that matches one of the VersaTiles patterns.

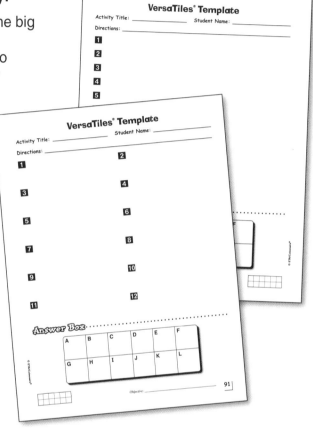

Quick Tip

Actually, there are more VersaTiles patterns and pattern codes than you might imagine. If you have used all the patterns and pattern codes provided on pages 92–93, try making up some of your own patterns and pattern codes. You can do this by working backward. Put the tiles upside down in the top part of the case and make any pattern you like. Then close the case and flip it over to figure out which number goes with which letter (the pattern code). You have a new VersaTiles pattern and pattern code! It's fun and rewarding!

Have Students Write VersaTiles® Activities

Students find that writing their own VersaTiles activities can be both fun and challenging. You might want to divide the class into teams and try this as a competition after students have become familiar with VersaTiles.

> You don't understand anything until you learn it more than one way.
>
> —*Marvin Minsky*

Before they begin, walk them through the process of writing an activity and educate them on the use of good distractors. As with anything else, you will find that students get better at writing VersaTiles activities with practice.

Students will need copies of the VersaTiles Templates (pages 90–91) and the sample Patterns and Pattern Codes (pages 92–93) to get started creating their own VersaTiles activities.

Writing VersaTiles activities is an ideal task for advanced students or for use as an extra credit project. You will find that some students enjoy the challenge; it's just like designing a puzzle!

After students become comfortable with writing their own activities, they can create new templates and pattern codes of their own. There are hundreds of possibilities!

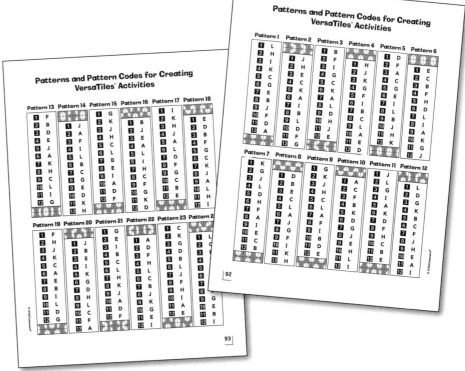

Patterns, Functions, and Algebra Mascot

Scope and Sequence Overview

A detailed Scope and Sequence for the VersaTiles® Mathematics Program is provided on pages 34–58. It is intended to help you assign meaningful practice to your students and to help you integrate VersaTiles into your current curriculum.

The Scope and Sequence is divided into categories according to the five strands in the program and the problem-solving activities are referenced separately for easy use.

Strand	Pages
Number and Number Concepts	34–37
Estimation and Computation	38–46
Patterns, Functions, and Algebra	47–49
Geometry and Measurement	50–53
Statistics and Probability	54–55
Problem-Solving Activities	56–58

Here's how to read the Scope and Sequence:

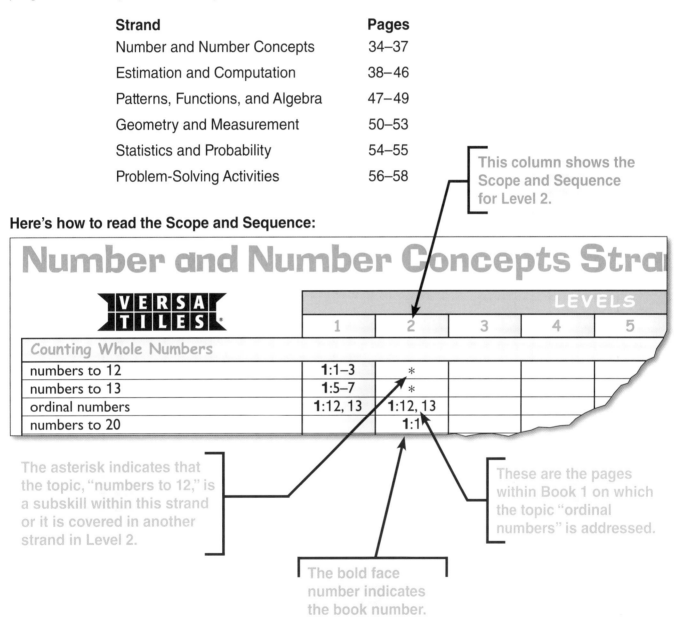

This column shows the Scope and Sequence for Level 2.

Number and Number Concepts Stra

| | LEVELS | | | | |
	1	2	3	4	5
Counting Whole Numbers					
numbers to 12	1:1–3	*			
numbers to 13	1:5–7	*			
ordinal numbers	1:12, 13	1:12, 13			
numbers to 20		1:1			

The asterisk indicates that the topic, "numbers to 12," is a subskill within this strand or it is covered in another strand in Level 2.

These are the pages within Book 1 on which the topic "ordinal numbers" is addressed.

The bold face number indicates the book number.

As you are perusing the Scope and Sequence, if it appears that a topic is missing within one strand at a specific grade level, check the other strands. You will find that skills (indicated by an asterisk) are integrated across strands when doing so is appropriate.

Number and Number Concepts Strand

VERSA TILES®

	Levels							
	1	2	3	4	5	6	7	8
Counting Whole Numbers								
numbers to 12	1:1–3							
numbers to 13	1:5–7	*						
ordinal numbers	1:12, 13	1:12, 13						
numbers to 20		1:1						
place-value models to 100	1:14, 16, 17	1:2, 4						
place-value models to 1,000		1:14, 15	1:2					
numbers to 1,000		1:18						
Naming Whole Numbers								
place-value models to 12	1:4	*						
place-value models to 20	1:18	*						
place-value models to 50	1:19, 20	1:6						
numbers to 50	1:22, 23	*						
place-value models to 100	1:15, 25, 26	1:8						
numbers to 100	1:27, 28, 31, 32	*						
place-value models to 1,000		1:16, 17						
numbers to 1,000		1:20, 21, 25, 26	1:14					
numbers to 100,000			1:15					
numbers through hundred thousands			1:9	1:12	1:1			
numbers through hundred billions					1:2			
numbers using scientific notation						1:3	1:7	
word names for numbers		1:3, 9, 19	1:2	1:3	1:1, 2	1:1		1:26

* This skill is a subskill within this strand or it is addressed in another strand at this level.

Number and Number Concepts Strand

VERSA TILES

	Levels							
	1	2	3	4	5	6	7	8
Comparing and Ordering Whole Numbers								
numbers to 13	1:8, 9							
numbers to 50	1:21, 24	1:7						
numbers to 100	1:29, 30	*						
numbers to 1,000		1:24						
numbers through hundred thousands			1:10, 11	1:9				
numbers through millions				1:7, 8				
numbers through hundred billions					1:5, 6	1:4	1:3	
Identifying Whole Numbers								
odd and even numbers		1:10, 11	1:1					
prime and composite numbers				*	1:18	1:14	1:11	1:5
rational and irrational numbers				*				1:30
Naming Fractions and Mixed Numbers								
fractional regions		1:27–29, 32						
unit fractions		1:27–29, 32	1:16					
non-unit fractions		1:27–29, 32	1:17, 18	1:13, 14	1:16			
mixed numbers			1:25, 26	1:21	1:22			
fractions and mixed numbers				1:13, 14, 21	1:16, 22	1:10	*	*
fractions in simplest form					1:21	1:13	1:14	1:8
equivalent fractions			1:20	1:15–17	1:17	1:11	*	*
Comparing and Ordering Fractions and Mixed Numbers								
fractions			1:19, 21, 24	1:18–20	*	*	*	*
mixed numbers				1:23	*	*	*	*
fractions and mixed numbers				1:18–20, 23	1:24	1:18	*	*
Naming Decimals								
in tenths			1:27	1:24	1:7			
in hundredths			1:28	1:25	1:7			
using scientific notation						1:6	1:32	1:27

* This skill is a subskill within this strand or it is addressed in another strand at this level.

Number and Number Concepts Strand

VERSA TILES

		Levels							
	1	2	3	4	5	6	7	8
Comparing and Ordering Decimals								
through hundredths			1:31, 32	1:29, 32	*	*	*	*
through thousandths					1:12, 13	1:8, 9	1:4	*
Naming Percents								
using models					1:30	1:29	*	*
Naming Ratios								
ratios						1:24	*	*
ratios in simplest form						1:25	1:21	1:18
equal ratios to form proportions						1:28	1:22	1:19
Naming Integers								
integers						1:32	1:29	1:22
classifying numbers								1:32
Identifying Place-Value								
3-digit numbers		1:22, 23	1:3–5	*				
place-value of 4-digit numbers			1:6, 7	1:1, 2				
5- or 6-digit numbers			1:8	1:4				
7-digit numbers				1:5, 6				
numbers through hundred billions					1:3	1:1		
decimals through thousandths				1:26	1:9	1:5		
equivalent forms of whole numbers		1:5	1:3, 5	1:3				
equivalent forms of decimals					1:10	1:7		
Rounding Whole Numbers, Fractions, and Decimals								
whole numbers through millions					1:4	*	1:1	*
decimals through ten thousandths					1:11	*	1:2	1:1
fractions to 0, $\frac{1}{2}$ or the next whole number					*	*	*	1:9

* This skill is a subskill within this strand or it is addressed in another strand at this level.

Number and Number Concepts Strand

VERSA TILES

	Levels							
	1	2	3	4	5	6	7	8
Comparing and Ordering Rational Numbers								
whole numbers, fractions, and decimals					1:27	1:21	*	*
fractions, decimals, and percents							1:28	1:21
decimals					*	*		1:2
integers							1:30	1:14
whole numbers, decimals, and integers							1:31	1:15
rational numbers								1:13, 31
Renaming Rational Numbers								
improper fractions as mixed numbers				1:22	*	*	1:15	*
mixed numbers as fractions							1:16	*
fractions as decimals			1:29	1:27	1:26	1:19	1:18	*
fractions as repeating decimals						1:20	1:19	1:11
fractions as percents					1:31	1:31		*
mixed numbers as decimals			1:30	1:28	*			*
decimals as fractions using models					1:7, 8			
decimals as fractions					1:25	*	*	*
decimals as rational numbers						1:19	1:17	1:10
decimals as percents					1:32	1:30	1:24	*
percents as fractions					1:31	1:31	1:23	1:20
percents as decimals					1:32	1:30	1:24, 25	1:20
Number Theory								
greatest common factor					1:20	1:12	1:13	1:7
least common multiple					1:23	1:17	1:20	1:12
prime factors					1:19	1:15		
prime factorization						1:2, 16	1:12	1:6
absolute value								1:23
square root					*			1:28, 29
numbers with exponents							1:6	*
numerical expressions without exponents						*	1:5	*
numerical expressions with exponents						*	1:10	1:3, 4

* This skill is a subskill within this strand or it is addressed in another strand at this level.

Estimation and Computation Strand

VERSA TILES

	Levels 1	2	3	4	5	6	7	8
Estimating Measurements								
length	2:1–6, 14–15							
weight	2:7–10, 13							
capacity	2:11–12							
Estimating Sums								
whole numbers		4:6, 10	4:3	2:11	2:1–6	2:1–4	*	*
decimals				2:26	2:23	3:1	2:1	2:1
fractions				5:21	5:11	4:1, 7	3:1	3:1, 7
Estimating Differences								
whole numbers		4:20, 24	4:13	2:12	2:8–11	2:8–11	2:2	*
decimals				2:26	2:24	3:2	2:2	2:2
fractions				5:26	5:14	4:8	3:2	3:2, 8
Estimating Products								
whole numbers			5:5, 6	3:14, 24, 25	3:2, 3, 13, 14	2:16, 20	*	*
decimals						3:9, 14	2:11, 14	2:7, 10, 11
fractions					5:22	4:17	3:18	3:15
Estimating Quotients								
whole numbers				4:9, 22	4:2, 3, 20, 21	2:23, 27	*	*
decimals						3:20, 23	2:22, 25	2:19, 22, 23
Using Models of Addition and Subtraction								
groups of objects	2:16–18, 20, 21							
pictographs	2:19, 22							
counting totals	2:24; 3:1, 2	2:1, 2						
counting the number being taken away and the number left	4:1–3	2:16, 17						
Skip Counting to Find Sums								
doubles	2:23							
skip count by 2	2:25, 26	3:9						
skip count by 3	2:27, 28	3:9						
skip count by 4	2:29, 30	3:12						
skip count by 5	2:31, 32	3:12						

* This skill is a subskill within this strand or it is addressed in another strand at this level.

38

VERSA TILES*

	Levels							
	1	2	3	4	5	6	7	8
Part–Part–Total Models								
sums	3:3	2:3						
addends	3:4; 4:4	2:4						
addition sentences	3:5–8	2:5						
subtraction sentences	4:5, 6, 8	2:18						
Addition Facts to 18								
adding 0–5	3:9–13							
using doubles	3:15	2:7, 8	2:4–6					
using words	3:25							
vertical form	3:28–30	2:14, 31, 32						
three addends	3:23, 24, 31, 32	2:11, 14	2:15	2:4				
using a number line			2:1–3					
using making a ten	3:24	2:11	2:7–9					
addition expressions	3:14	2:6						
addition facts	3:18, 19	2:9, 10						
addition sentences	3:16, 17, 20	*	*					
missing addends and sums	3:21, 22	2:9, 15, 30	2:10–15	2:1, 3, 16				
order property			2:12, 13	2:16				
grouping property			2:14–15					
Addition of Whole Numbers								
tens		4:1	4:1	2:5				
hundreds		4:1	4:2	2:6				
1- and 2-digit numbers without regrouping		4:2, 3	4:4	2:13, 18				
3-digit numbers without regrouping		4:4	*	2:13				
1- and 2-digit numbers with regrouping		4:8, 9, 31	4:5, 6	*				
3- and 4-digit numbers with regrouping		4:12, 13, 31	4:7–10	2:7, 14, 15	2:14–16	2:5, 6		
5- to 7-digit numbers with regrouping					2:17	2:7		
money	*	4:32		2:29				
related addition and subtraction facts		4:30						
order and grouping properties				2:17				

* This skill is a subskill within this strand or it is addressed in another strand at this level.

Estimation and Computation Strand

VERSA TILES

	Levels							
	1	2	3	4	5	6	7	8
Addition of Decimals								
tenths			4:30	2:27	2:25	3:3	2:3	*
hundredths			4:31	2:28	2:26	3:4	2:4	2:3
thousandths					2:27	3:5	2:5	2:4
Addition of Fractions and Mixed Numbers								
fractions with like denominators				5:1–5	5:1, 2	4:2	3:3	3:3
fractions with unlike denominators				5:13, 14, 17, 20	5:3	4:3, 6	3:4, 7	3:4
mixed numbers with like denominators				5:22, 23	5:12	4:9	3:10	*
mixed numbers with unlike denominators				5:24, 25	5:13	4:10	3:11	3:9
Addition of Integers								
using counters							4:1–4	4:1–4
using a number line							4:5, 6	4:5, 6
using rules							4:7, 8, 17, 29, 30	4:7, 8, 16, 17, 27, 30
Subtraction Facts to 18								
subtracting 0–5	4:10–14							
using words	4:25							
vertical form	4:28–32	2:31, 32	*					
using a number line		2:21	2:16–22, 25	2:1				
using making a ten		2:21	2:23–25	2:1				
money		2:28–30						
subtraction expressions	4:15	2:19						
using addition facts	4:18, 19, 23, 24	2:24, 25; 4:30						
subtraction sentences	4:7, 9, 22	2:20	2:26, 27					
missing differences and subtrahends	4:20, 21	2:20; 4:30	2:26, 27	2:2, 3				

* This skill is a subskill within this strand or it is addressed in another strand at this level.

Estimation and Computation Strand

VERSA TILES

			Levels					
	1	2	3	4	5	6	7	8
Subtraction of Whole Numbers								
tens		4:16	4:11	2:5				
hundreds		4:16	4:12	2:6				
1- and 2-digit numbers without regrouping		4:17, 18	4:14	2:18				
3-digit numbers without regrouping		4:19						
1- and 2-digit numbers with regrouping		4:22, 23	4:15, 16	2:7, 19				
3- and 4-digit numbers with regrouping		4:26, 27, 31	4:17–23	2:20, 21, 24	2:18, 19	2:12, 13		
5- to 7-digit numbers with regrouping					2:22	2:14		
money		4:32	4:25					
using related facts		4:30						
using equal additions					2:7			
Subtraction of Decimals								
tenths			4:30	2:27	2:28	3:6	2:8	*
hundredths			4:32	2:28, 32	2:29	3:7	2:9	2:5
thousandths					2:32	3:8	2:10	2:6
Subtractions of Fractions and Mixed Numbers								
fractions with like denominators				5:6–8	5:5	4:4	3:5	3:5
fractions with unlike denominators				5:15, 16, 17, 20	5:4, 6	4:5, 6	3:6, 7	3:6
mixed numbers with like denominators				5:27, 28	5:15	4:11	3:12	*
mixed numbers with unlike denominators				5:29, 32	5:16, 17	4:12, 13	3:13, 14	3:10, 11
Subtractions of Integers								
using counters							4:9–12	4:9–11
using a number line							4:13, 14	4:12, 13
using rules							4:15–17, 29, 30	4:14–17, 27, 30

* This skill is a subskill within this strand or it is addressed in another strand at this level.

Estimation and Computation Strand

VERSA TILES

	Levels							
	1	2	3	4	5	6	7	8
Models of Multiplication and Division								
number of equal groups		3:1–4, 18, 19, 26, 27						
number in each equal group		3:20, 21, 25, 27						
total		3:22, 23						
number left over (remainder)		3:24						
number sentences			3:1, 18					
Multiplication Facts								
factors of 0 or 1		3:14, 32	3:3, 4	3:1				
factors of 2 or 3		3:10, 11, 32	3:3–6	3:1, 3				
factors of 4 or 5		3:13, 32	3:3–6	3:1, 2				
factors of 6 or 7		3:32	3:7, 8	3:2, 4				
factors of 8 or 9		3:32	3:9, 10	3:3, 4				
using skip counting		*	3:2					
three factors			3:14, 15	3:6				
multiplication facts		3:5–8, 15	3:11	*				
missing factors			3:11–13	3:5, 8, 9				
order and grouping properties			3:12, 13	3:8, 9				
Multiplication of Whole Numbers								
2- and 3-digit numbers by 1-digit numbers			5:1–4, 7–13, 16	3:12, 13, 15–18	3:5–7			
4- to 6-digit numbers by 1-digit numbers					3:1, 8–10	2:15, 17		
2- and 3-digit numbers by 2-digit numbers				3:22, 23, 26–29	3:15–17	*		
4- and 6-digit numbers by 2-digit numbers				3:32	3:18, 19	2:21, 22		
missing factors			*	*	3:11, 12			
distributive property				3:7	3:4			

* This skill is a subskill within this strand or it is addressed in another strand at this level.

Estimation and Computation Strand

VERSA TILES

	Levels							
	1	2	3	4	5	6	7	8
Multiplication of Decimals								
whole numbers				3:19	3:22, 23, 26	3:10, 11, 15, 16, 19	2:12, 13, 15, 16	2:8, 9, 12, 13
powers of 10					*	3:26	2:28	2:28
money				3:19				
decimals in tenths by decimals in tenths					3:27, 28, 31	3:17	2:17	2:16
decimals in hundredths by decimals in tenths					3:29–33	3:29, 32	2:18	2:17
decimals in hundredths by decimals in hundredths						3:18	2:18, 19	2:17, 18
Multiplication of Fractions								
area models					5:20, 23			
whole numbers					5:21	4:16	3:15	3:14
fractions by fractions					5:24, 25	4:18, 19	3:19, 20	3:16, 17
reciprocals					5:30	4:25	3:23, 24	3:21
fractions by mixed numbers					5:26	4:22	3:21	3:18
mixed numbers by mixed numbers					5:27	4:23, 24	3:22	3:19, 20
Multiplication of Integers								
using counters							4:20	4:20
using patterns							4:21	4:21
using rules							4:22, 23, 28–30	4:22, 27, 30
numbers sentences							*	4:26
Division Facts								
dividing by 0 or 1		3:32	3:19, 20	4:1				
dividing by 2 or 3		3:28, 32	3:19, 20–22	4:1, 3				
dividing by 4 or 5		3:29, 32	3:19, 20–22	4:1, 2				
dividing by 6 or 7		3:32	3:23, 24	4:2, 4				
dividing by 8 or 9		3:32	3:25–27	4:3, 4				

* This skill is a subskill within this strand or it is addressed in another strand at this level.

Estimation and Computation Strand

VERSA TILES

	Levels							
	1	2	3	4	5	6	7	8
Division of Whole Numbers								
using basic facts and patterns				4:20, 21	4:1, 18, 19			
area models					4:11			
2-digit numbers by 1-digit numbers without remainders			5:17, 18, 21–24, 32	4:10	4:4, 5			
3-digit numbers by 1-digit numbers without remainders			5:19, 20	4:8, 12, 13, 15–17	4:6, 7			
4- to 5-digit numbers by 1-digit numbers without remainders					4:10	2:25, 26		
money			5:32	4:16, 17, 26				
2-digit numbers by 1-digit numbers with remainders			5:25–29	4:7, 11				
3-digit numbers by 1-digit numbers with remainders						2:24		
2- to 3-digit numbers by 2-digit numbers with remainders				4:14, 23–25, 27	4:22–24	2:28		
4- to 5-digit numbers by 2-digit numbers with remainders					4:25	2:29, 32		
multiples of 10				4:23–26				
missing dividends, divisors, and quotients			3:30	4:5				
Division of Decimals								
dividing by whole numbers					4:12–14, 28–30	3:21, 22, 24, 25	2:23, 24, 26, 27	2:20, 21, 26, 27
fractions in tenths divided by fractions in tenths						3:27	2:29	2:29
fractions in hundredths divided by fractions in tenths							2:30	
fractions in tenths divided by fractions in hundredths								2:30
fractions in hundredths divided by fractions in hundredths						3:28, 29, 32	2:31, 32	2:30–32
rounding quotients					4:31, 32	*	*	*

* This skill is a subskill within this strand or it is addressed in another strand at this level.

Estimation and Computation Strand

	Levels							
	1	2	3	4	5	6	7	8
Division of Fractions								
whole numbers divided by fractions					5:28, 29	4:26–28	*	*
reciprocals					5:31, 32	4:25	3:25, 28	3:24–26
fractions divided by mixed numbers						4:30	*	*
mixed numbers divided by fractions						4:29	3:29, 30	3:27, 28
mixed numbers divided by mixed numbers						4:31, 32	3:31, 32	3:29–32
Division of Integers								
using multiplication							4:26	*
using inverse operations								4:23, 24
using rules							4:27–30	4:25, 27, 30
number sentences							*	4:26
Fact Families								
addition and subtraction		2:26, 27	2:32					
multiplication and division			3:31, 32					
Regrouping, Rounding, and Renaming Numbers								
regrouping 2- and 3-digit numbers		4:7, 11, 21, 25						
rounding whole numbers			*	2:8–10	*	*	*	*
rounding decimals		4:5		2:25	4:15	*	*	*
rounding fractions					5:7, 10	*	*	*
renaming decimals in scientific notation				4:31, 32			4:31, 32	4:31, 32
Number Theory								
factors				3:10	*	*	*	*
multiples				5:9–12	*	*	*	*
averages				4:30	*	*	*	*
divisibility rules (2, 5, 10)				4:31, 32				
prime numbers				3:11	*	*	*	*

* This skill is a subskill within this strand or it is addressed in another strand at this level.

Estimation and Computation Strand

		Levels							
		1	2	3	4	5	6	7	8
Ratios and Proportions									
ratios							5:1–5	5:1–4	5:1–4
equal ratios							5:6, 7	5:5, 6	5:5, 6
rates							5:8	5:7	5:7
unit rates							5:9	5:8	5:8
unit prices							5:10, 11	5:9, 10	5:9, 10
proportions							5:14, 16–18	5:11, 5 13–15	5:11, 13–15
cross products							5:15	5:12	5:12
scale length							5:19–21	5:16, 17	5:16, 17
Percents									
10 by 10 grids					*	*	5:24–26	*	*
renaming percents as fractions							*	5:20, 21	5:20, 21
percent of a number							5:27–29, 32	5:22–25	5:22–25
percents of increase and decrease								5:26, 27	5:26, 27
sales tax								5:28	5:28
discounts and sale prices								5:29	5:29
simple interest								5:32	5:32

* This skill is a subskill within this strand or it is addressed in another strand at this level.

Patterns, Functions, and Algebra Strand

	1	2	3	4	5	6	7	8
Shape Patterns								
real-life objects	5:1, 4							
2- and 3-dimensional shapes	5:6–8	5:3–5	6:29, 30	6:28–30	6:30			
stacked cubes			6:28	6:25	6:31			
slides, flips, and turns			6:31, 32	6:31, 32	*	*	*	*
tessellations					6:29			
number cubes					6:32			
Identifying Shapes								
similar objects	5:2–3	*						
similar geometric shapes	5:5	5:1, 2						
different shapes	*	5:6, 7						
Number Patterns								
counting forward and backward	5:9, 10	5:9						
repeating patterns	5:21, 22	5:20, 21	6:13–15					
odd and even numbers		5:8	6:7	6:7				
sequences			6:7	6:14, 15	6:26–28	6:7, 8	6:4, 5	6:3, 4
Skip Counting Patterns								
skip counting by 2 to 100	5:11, 12	5:10						
skip counting by 3 to 100		5:11						
skip counting by 5 to 100	5:13, 14							
skip counting by 10 to 100	5:15, 16	5:12						
skip counting by 10 to 1,000		5:13						
skip counting by 2 through 9			6:5	6:5				
skip counting by 10; 100; 1,000; or 10,000	*	*	6:6, 8	6:6, 8				
multiplication patterns by 10; 100; or 1,000				6:13				

Levels

VERSA TILES®

* This skill is a subskill within this strand or it is addressed in another strand at this level.

Patterns, Functions, and Algebra Strand

VERSA TILES®

	Levels							
	1	2	3	4	5	6	7	8
Money Patterns								
increases and decreases by 5¢, 10¢, 25¢, 50¢, or $1.00			6:1, 2	6:1, 2				
Time Patterns								
increases and decreases by the same time interval			6:3, 4	6:3, 4				
Addition and Subtraction Patterns								
addition facts	5:23–26	*						
subtraction facts	5:29–32	*						
addition sentences	*	5:22–25						
subtraction sentences	*	5:28–30						
Addition and Multiplication Tables								
addition tables	5:27, 28	5:26, 27	6:11	6:11				
multiplication tables		5:31, 32	6:12	6:12				
missing addends		5:18	6:21	6:20				
missing factors		5:19	6:22	6:21				
magic squares			6:9, 10	6:9, 10				
Function Machines								
outputs (sums, differences, products, quotients)	5:17	5:14	6:18, 19	6:19	6:19, 20	*	*	6:29
inputs (addends, factors)	5:18	5:15	*	6:19	6:19, 20	*	*	6:29
rules	5:19, 20	5:16, 17	6:20	6:18	*	*	*	6:29
Expressions								
order of operations					6:1, 2			
addition					6:3, 4			
subtraction					6:5, 8			
addition, subtraction, multiplication, division						6:5, 6	6:2, 3	6:2
naming expressions						6:1–4	6:1	6:1

* This skill is a subskill within this strand or it is addressed in another strand at this level.

Patterns, Functions, and Algebra Strand

VERSA TILES®

	Levels							
	1	2	3	4	5	6	7	8
Equations								
models of addition equations					6:9			
one-step addition equations					6:10, 13	6:13–15	6:7–9, 22	6:10
models of subtraction equations					6:11			
one-step subtraction equations					6:12, 16	6:16–18	6:10–12	6:11
one-step addition and subtraction equations						6:32	6:23	6:6, 7
models of multiplication equations					6:17			
one-step multiplication equations					6:18	6:19–21	6:13–15, 24	6:12
one-step division equations					6:21	6:24, 25, 28	6:16–18, 25	6:13
one-step multiplication and division equations					6:22	*	*	6:8, 9
one-step equations					*	*	6:19	*
two-step equations						6:9, 10		6:14–17
naming equations						6:11, 12, 31	6:6	6:5
substitutions							*	*
systems of equations								6:32
Graphing								
naming points			6:25	6:24	6:25	6:29	6:26	6:22
naming coordinates			6:23, 24	6:22, 23	6:23, 24	6:30	6:27	6:22
equations							6:32	6:30
slope								6:31
Algebra								
function tables							6:28–29	6:28
Pythagorean theorem								6:20, 21, 23
sine, cosine, and tangent								6:24, 25

* This skill is a subskill within this strand or it is addressed in another strand at this level.

Geometry and Measurement Strand

VERSA TILES

	1	2	3	4	5	6	7	8
Spatial Relationships								
above or below	6:1, 2							
left or right	6:3, 4							
inside or outside		6:1						
views of stacked cubes				7:13				
Identifying 2- and 3-dimensional Shapes								
descriptions of 2- and 3-dimensional shapes	6:8–13	6:6–9						
3-dimensional shapes			7:24, 25	7:15	7:28			
2-dimensional shapes			7:26, 27	7:16, 17	7:4			
geometric figures				7:19, 20	7:1			
triangles					7:5	7:5	7:5	7:6
circles					7:6			
quadrilaterals						7:5	7:5	7:6
Plane Figures								
lines, line segments, rays, and angles			7:30	7:18	7:1	7:1	7:1	7:1
intersecting, perpendicular, and parallel lines						7:2	7:2	7:2, 5
lines of symmetry		6:13	7:31, 32	7:23, 24	*	7:10	7:10	7:10
classifying angles					7:3	7:3	7:3	7:3
Congruencies and Similarities								
matching groups of objects	6:5	6:2						
congruent 3-dimensional figures	6:6, 7	6:4	7:24	7:14, 21, 22	7:7	7:6	7:6	7:7
congruent 2-dimensional shapes	6:6, 7, 16, 17	6:5, 10–12	7:26	7:21, 22	7:7	7:6	7:6	7:7
similar figures				7:21, 22	*	7:7	7:7	*
Transformations								
translations					7:8	7:8	7:8	7:8
reflections					7:8	7:8	7:8	7:9
rotations					7:9	7:9	7:9	*

Levels

* This skill is a subskill within this strand or it is addressed in another strand at this level.

Geometry and Measurement Strand

VERSA TILES®

	Levels 1	2	3	4	5	6	7	8
Distance Around Figures								
perimeter		6:20	7:9–11	7:25	7:20, 21	7:17	7:18	7:13
circumference					7:26	7:26	7:23	7:19
Area								
counting square units		6:21, 22	7:12	7:28				
squares and rectangles		6:21, 22	7:12	7:29	7:22, 23	7:20	7:19	7:16
parallelograms					7:25	7:23	7:22	7:16
triangles					7:24	7:21, 22	7:20, 21	7:17
trapezoids								7:18
circles					7:27	7:27	7:24	7:20
irregular figures						7:28	7:25	7:21
Surface Area								
rectangular prisms							7:28	7:24
cylinders							7:29	7:25
Volume								
counting cubes		6:3	7:13	7:32	7:29			
cubes and rectangular prisms				7:32	7:30	7:29	7:30	7:26
triangular prisms						7:30	7:31	7:26
cylinders						7:31	7:32	7:27
pyramids								7:28
cones								7:29
irregular figures						7:32		7:32
Scale								
finding actual distances				7:26, 27	7:31			
finding scale distances					7:32			

* This skill is a subskill within this strand or it is addressed in another strand at this level.

Geometry and Measurement Strand

	Levels 1	2	3	4	5	6	7	8
Time								
more/less time	7:1, 2	7:1						
telling time to the hour	7:3–7	7:2						
telling time to the half-hour	7:8–12	7:4–6						
telling time to the quarter-hour		7:7–9						
telling time to the nearest five minutes		7:10, 11						
estimating		7:12						
a.m. and p.m.		7:3						
Calendar								
days and dates	7:13, 14	7:13, 14						
months	7:15							
Money								
counting coins	7:18–25	7:15, 18, 19						
counting bills and coins		7:21, 22						
buying items	7:26, 27	7:27, 28						
comparing and ordering	7:28, 29, 32	7:32						
finding equal values of coins		7:20, 23						
sums and differences		7:24–26						
calculating change		7:29						
Temperature								
Fahrenheit thermometer	6:30	6:30						
estimating	6:31, 32	6:31						
Estimating Length, Weight, Capacity, and Mass								
length in customary units	6:22	6:18	7:3, 4	*	*	7:11		
length in metric units			7:7, 8	*	*	7:14		
weight	6:24	6:23	7:18, 19	7:7	7:13	7:12		
mass	6:28	6:24	7:22, 23	7:10	7:16	7:15		
capacity in customary units	6:25	6:25	7:16, 17	7:9	7:13	7:12		
capacity in metric units	6:25, 29	6:28	7:20, 21	7:11	7:16	7:15		
customary measurement sense		6:29						
metric measurement sense		6:32						

* This skill is a subskill within this strand or it is addressed in another strand at this level.

Geometry and Measurement Strand

VERSA TILES®

	1	2	3	4	5	6	7	8
Measuring Length								
nonstandard units	6:18, 19	6:16						
customary units	6:20, 21	6:17	7:1, 2	7:1–3	7:12			
metric units	6:23	6:19	7:5, 6	7:5	7:15			
Measuring Angles								
using an angle ruler or protractor					7:2	7:4	7:4	7:4
Conversion of Units								
length in customary units				7:4, 8	7:14	7:13	7:11	7:11
length in metric units				7:6	7:17	7:16	7:13	7:12
weight				7:8	7:14	7:13	7:12	7:11
mass				7:12	7:17	7:16	7:14	7:12
capacity in customary units				7:8	7:14	7:13	7:12	7:11
capacity in metric units				7:12	7:17	7:16	7:14	7:12
units of time							7:15	
Triangles								
identifying types of					7:5	7:5	7:5	7:6
Pythagorean theorem								6:20, 21, 23
sine, cosine, and tangent								6:24, 25

Levels

* This skill is a subskill within this strand or it is addressed in another strand at this level.

Statistics and Probability Strand

	1	2	3	4	5	6	7	8
Gathering Information								
context of a story	8:1, 2	8:1, 2						
sensible answers	8:32							
Sorting Information								
objects by category	8:3–6	8:3, 4						
positions of inside, outside, and on	8:7							
Venn diagrams	8:8	8:5, 6						
Reading and Interpreting Tables								
tally marks	8:9, 10	8:7, 8	8:6–9					
tables	8:11–13	8:9, 10	8:1–4	8:1, 3, 4	8:1, 3	8:1	8:1	8:1, 2
schedules	8:14, 15	8:11, 12	8:5	8:2	8:2			
Reading and Interpreting Graphs								
real object graphs	8:16, 17	8:13, 14						
pictographs	8:18–21	8:15–18	8:14–16	8:16, 17	*	8:2, 3	8:2	
single vertical bar graphs	8:22, 23	8:19, 20	8:17, 18		8:15	8:4	8:3	
double vertical bar graphs					8:22–23	8:5	8:4	8:3
single horizontal bar graphs	8:24, 25	8:21, 22	8:19–22	8:18, 19	8:14			
double horizontal bar graphs						8:6	8:5	8:4
circle graphs				8:24, 25	8:20, 21	8:9, 10	8:9, 10	8:1, 14
timelines			8:10–11	8:14–15				
single line graphs			8:12, 13	8:20–23	8:18, 19	8:7	8:6	
double line graphs						8:8	8:7	8:7, 8
line plots				8:13	8:12, 13		8:8	8:9, 10
stem-and-leaf plots					8:24	8:18, 19	8:18	8:15, 16
histograms								8:5, 6
other graphs						8:20, 21	8:16–17	8:12–13
misleading graphs								8:30, 31
Analyzing Data								
range				8:5, 6	8:4, 5	8:11, 12	8:11	8:17
mode				8:7, 8	8:6, 7	8:13	8:12	8:18
median				8:9, 10	8:10, 11	8:16, 17	8:14, 15	8:20
mean				8:11, 12	8:8, 9	8:14, 15	8:13	8:19

Levels

Statistics and Probability Strand

VERSA TILES

	Levels							
	1	2	3	4	5	6	7	8
Certain, Possible, and Impossible Events								
likely and unlikely events	8:28, 29	8:23, 24						
real and not real events	8:30, 31							
certain and impossible events		8:25, 26						
Listing Outcomes								
spinners			8:23, 24	8:26	8:25	8:24, 25	8:19	8:21
groups of cards		8:30	8:25, 26	8:27			8:19	8:21
number cubes						8:25	8:19	8:21
coins		8:27				8:24, 25	8:19	8:21
counting principle							8:20	
real life situations								8:22
Independent/Dependent Events								
spinning a spinner			8:27–29	8:28	8:28	8:29	8:21	8:25
picking a card from a group of cards			8:30–32	8:29	8:29	8:28, 29	8:23	
rolling a number cube					8:30	8:29	8:22	
independent events						8:30	8:28	8:26, 27, 29
dependent events								8:28, 29
Making Predictions								
spinner outcomes in a given amount of spins				8:30	8:31	8:31	8:24	
card outcomes in a given number of picks				8:31				
number cube outcomes in a given number of rolls					8:32	8:32	8:25	
sample surveys								8:32
Combinations and Permutations								
combinations		8:31, 32		8:32			8:31, 32	8:24
permutations							8:29, 30	8:23

* This skill is a subskill within this strand or it is addressed in another strand at this level.

Problem-Solving Activities

VERSA TILES.

	Levels							
	1	2	3	4	5	6	7	8
Problem Solving: Using Information from Pictures and Charts								
tables	7:30–31	7:30–31	1:12–13	1:10–11; 3:20–21	1:14–15; 3:24–25	1:22–23; 1:26–27; 2:18–19	1:8–9; 2:6–7	1:16–17
tables of paired values			6:26–27					
pictures	6:14–15	6:14–15			2:20–21			
drawings					7:10–11			7:30–31
lists			3:16–17			3:12–13	1:26–27	1:24–25
groups of objects	8:26–27	8:28–29						
fractional amounts		1:30–31	1:22–23					
length of objects	2:14–15							
diagrams			6:16–17	6:26–27	5:8–9; 8:16–17	7:24–25; 8:22–23	7:26–27	2:14–15; 7:14–15, 22–23
bar graphs	1:10–11		8:20–21					
pictographs	7:16–17	7:16–17			8:22–23			
timelines			8:10–11	8:14–15 8:22–23				
line graphs								
circle graphs						5:30–31	5:30–31	5:30–31
maps							3:8–9	
various graphs							8:16–17	8:12–13
misleading graphs								8:30–31
Problem Solving: Using Number Sense Skills								
using logical reasoning			2:30–31; 7:14–15	2:30–31	1:28–29			
extending patterns			7:28–29					
needed and extra information						4:14–15	3:16–17	3:12–13
estimation					2:12–13			
quotients and remainders			5:30–31	4:18–19		2:30–31		
working backward					6:14–15	4:20–21	3:26–27; 6:30–31	3:22–23
decimals				1:30–31				

* This skill is a subskill within this strand or it is addressed in another strand at this level.

Problem-Solving Activities

VERSA TILES

	Levels							
	1	2	3	4	5	6	7	8
Problem Solving: Using Geometry and Algebra								
perimeter and area formulas				6:16–17; 7:30–31				
using given formulas	6:26–27					6:22–23	6:20–21	6:18–19
units of measure		6:26–27			7:18–19	7:18–19		
conversion of units							7:16–17	
substituting values into formulas					6:6–7			
Pythagorean theorem								6:26–27
Problem Solving: Using the Four Operations								
addition	3:26–27; 4:26–27	2:12–13; 4:14–15		5:18–19, 30–31	4:16–17			
subtraction	4:26–27	2:22–23; 4:28–29		5:18–19, 30–31	4:16–17			
multiplication					4:16–17			
division					4:8–9, 16–17			
Problem Solving: Multi-Step Problems								
whole numbers			5:14–15		3:20–21	2:18–19; 6:26–27		
whole numbers and decimals			4:28–29		2:30–31	3:30–31		2:24–25
whole numbers, decimals, and fractions					5:18–19; 8:26–27			
decimals							2:20–21	
fractions							8:26–27	
integers							4:18–19, 24–25	4:18–19, 28–29
averages					4:26–27			
rates						5:12–13	5:18–19	5:18–19
proportions						5:22–23		
using a sign in/sign out sheet						8:26–27		

* *This skill is a subskill within this strand or it is addressed in another strand at this level.*

Problem-Solving Activities

VERSA TILES

Problem Solving: Using Number Sentences and Equations	Levels							
	1	2	3	4	5	6	7	8
addition sentences			2:28–29; 3:28–29; 4:26–27	2:22–23; 3:30–31; 4:28–29				
subtraction sentences	4:16–17		2:28–29; 3:28–29; 4:26–27	2:22–23; 3:30–31; 4:28–29		3:30–31		
multiplication sentences		3:16–17	3:28–29	3:30–31; 4:28–29				
division sentences		3:30–31	3:28–29	4:28–29				

* This skill is a subskill within this strand or it is addressed in another strand at this level.

How to Use VersaTiles®

Place  in front of you.

Open it.

Place all of the number tiles (1–12) in order, in the lid.

 Open your book. Read the directions.

Start with .

Pick up , which corresponds to .

Read the question. Think about it. Find the answer in the Answer Box.

Find the letter of the correct answer in the Answer Box. Place over the same letter in the Answer Case.

Repeat this for each tile, until they have all been placed in the Answer Case.

Close the lid.  Flip it over.  Re-open it.

Look at the colorful pattern! Compare the pattern to the one in your book. If it matches, you have completed the entire activity correctly! If not, turn over only the tiles that do not match. Flip the case back over, and redo the exercises that are incorrect. Close the Answer Case again and recheck. When the patterns match exactly you have successfully completed the activity!

VersaTiles® Home Lending Library Tracking Form #1

Book Number: _____

Book Title: _____

Date	Student Name	Parent/Guardian Signature	Notes to the Teacher*

*Please use a separate sheet of paper if more space is required. Thank you!

VersaTiles® Home Lending Library Tracking Form #2

Student Name: _____

Book Number	Book Title	Packet Number	Date Out	Date In	Circle the pages you completed.
1	Counting 1-100				1 2 3 4 5 6 7 8 9 10 11 12 13 14 15 16 17 18 19 20 21 22 23 24 25 26 27 28 29 30 31 32
2	Let's ESTIMATE				1 2 3 4 5 6 7 8 9 10 11 12 13 14 15 16 17 18 19 20 21 22 23 24 25 26 27 28 29 30 31 32
3	Let's ADD				1 2 3 4 5 6 7 8 9 10 11 12 13 14 15 16 17 18 19 20 21 22 23 24 25 26 27 28 29 30 31 32
4	Let's SUBTRACT				1 2 3 4 5 6 7 8 9 10 11 12 13 14 15 16 17 18 19 20 21 22 23 24 25 26 27 28 29 30 31 32
5	All About PATTERNS				1 2 3 4 5 6 7 8 9 10 11 12 13 14 15 16 17 18 19 20 21 22 23 24 25 26 27 28 29 30 31 32
6	All About SHAPES and MEASURES				1 2 3 4 5 6 7 8 9 10 11 12 13 14 15 16 17 18 19 20 21 22 23 24 25 26 27 28 29 30 31 32
7	TIME and MONEY				1 2 3 4 5 6 7 8 9 10 11 12 13 14 15 16 17 18 19 20 21 22 23 24 25 26 27 28 29 30 31 32
8	TALLIES and TABLES				1 2 3 4 5 6 7 8 9 10 11 12 13 14 15 16 17 18 19 20 21 22 23 24 25 26 27 28 29 30 31 32

Home Lending Library Wall Chart

Packet Number	Student Name	Date Out	Date In	Notes

Observation and Interview Guide

Here are examples of questions you might ask a student to find out what he/she was thinking about while working on the benchmark activities. Choose questions from the list that may apply, or create your own questions to evaluate the student's knowledge and to determine exactly what constitutes purposeful practice for that student.

Mathematical Reasoning Skills

- What is this activity about?
- Would you explain to me in your own words how you did this activity?
- What strategies or procedures did you use to complete this activity?
- How do you know that your answers are correct?
- In what real-world situation might you use the skills addressed in this activity?

Mathematical Tools and Techniques

- What steps did you take to complete this activity?
- Did you record your work?
- What tools did you use to do it (diagrams, sketches, manipulatives, calculators, and so on)?
- Have you tried making a guess?
- Can the Answer Box help you get started?
- Is there a hint? If so, can it help you get started?
- Have you ever done exercises or problems like the ones presented in this activity? If so, how did you solve them?

Mathematical Understanding

- What were the mathematical ideas in this activity?
- What mathematics did you learn by doing this activity?
- Did you discover a new way of doing a mathematical procedure?
- How are the ideas and procedures you used in this activity similar to other things we have done in class?

Mathematical Communication

- Would you explain this activity to me in your own words?
- How would you explain this activity to a younger student?
- Could you illustrate how you completed this activity by drawing a sketch or making a model?
- What were the key words or symbols in this activity? What do they mean?

Mathematical Dispositions

- What did you like most/least about this activity?
- What did you find difficult about this activity?
- What was easy about this activity?
- Are you confident that you could do an activity like this one on your own?
- Do you enjoy doing activities like this one?

Observation and Interview Notes

Student Name: _____ Date(s): _____

Benchmark Activity: _____ Level: _____ Book Number: _____

Interview Questions	Observation and Interview Notes
Mathematical Reasoning	
Mathematical Tools and Techniques	
Mathematical Understanding	
Mathematical Communication	
Mathematical Dispositions	

Benchmark Activities Answer Patterns

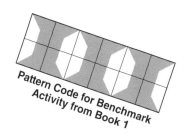

Pattern Code for Benchmark
Activity from Book 1

Pattern Code for Benchmark
Activity from Book 2

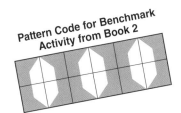

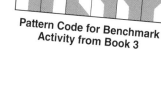

Pattern Code for Benchmark
Activity from Book 3

Pattern Code for Benchmark
Activity from Book 4

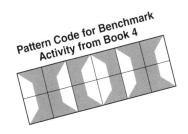

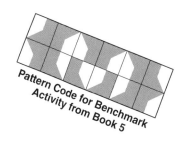

Pattern Code for Benchmark
Activity from Book 5

Pattern Code for Benchmark
Activity from Book 6

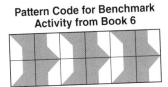

Pattern Code for Benchmark
Activity from Book 7

Pattern Code for Benchmark
Activity from Book 8

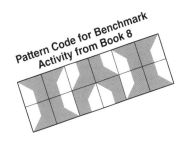

Which Comes Just After?

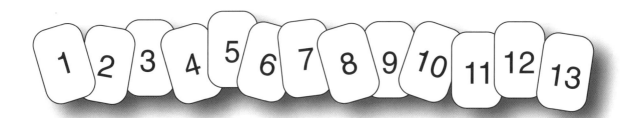

Which number comes just *after*?

1 2, 3, ▪

2 0, 1, ▪

3 8, 9, ▪

4 5, 6, ▪

5 10, 11, ▪

6 7, 8, ▪

7 4, 5, ▪

8 9, 10, ▪

9 1, 2, ▪

10 6, 7, ▪

11 3, 4, ▪

12 11, 12, ▪

6 is just after 5.

Answer Box . . . .

A	B	C	D	E	F
10	3	7	4	9	2
G	H	I	J	K	L
12	5	6	8	13	11

Objective: Name the third counting number to 13 in a
series of three numbers.

Many Pennies

Dan has 4 pennies.

My Pennies

Dan   Amy   Kate   Mike   Ryan   Otis   Suki

Each 🪙 stands for 1 penny.

How many pennies?

1 Amy

2 Kate

3 Mike

4 Ryan

5 Otis

6 Suki

Who has more?

7 Dan or Otis

8 Mike or Kate

9 Otis or Amy

10 Amy or Ryan

11 Suki or Mike

12 Dan or Suki

Answer Box

A	B	C	D	E	F
8	5	Otis	6	2	1
G	**H**	**I**	**J**	**K**	**L**
Kate	Suki	Mike	3	Amy	Dan

Objective: Identify how many and who has more based on information in a pictograph.

Adding Two!

Example

Use counters to find the sum.

2 + 2

2 + 2 = 4

Use counters to find the sum.

1 5 + 2

2 3 + 2

3 0 + 2

4 8 + 2

5 1 + 2

6 7 + 2

7 9 + 2

8 6 + 2

9 11 + 2

10 10 .

2 more .

in all.

11 4 .

2 more .

in all.

12 2 .

2 more .

in all.

Answer Box

A	B	C	D	E	F
11	12	5	4	3	6
G	H	I	J	K	L
7	10	8	9	13	2

Objective: Use counters to find the sum when adding 2.

More Words to Try

Find the missing number.

1

10	
■	4

10 minus ■ equals 4.

2

12	
5	■

12 minus 5 equals ■.

3

9	
9	■

9 minus 9 equals ■.

4

12	
2	■

12 minus 2 equals ■.

5

7	
5	■

7 minus 5 equals ■.

6

11	
■	8

11 minus ■ equals 8.

7

6	
5	■

6 minus 5 equals ■.

8

12	
1	■

12 minus 1 equals ■.

9

9	
■	1

9 minus ■ equals 1.

10

10	
■	5

10 minus ■ equals 5.

11

12	
■	3

12 minus ■ equals 3.

12

11	
■	7

11 minus ■ equals 7.

Answer Box

A	B	C	D	E	F
10	6	8	0	3	7
G	H	I	J	K	L
1	5	9	2	4	11

Objective: Complete a sentence that describes a part-part-total model.

Terrific Tens!

Find the next number in the pattern.

1 10, 20, 30, ■

2 40, 50, 60, ■

3 20, 30, 40, ■

4 60, 70, 80, ■

5 50, 60, 70, ■

Count by 10!

6 15, 25, 35, ■

7 45, 55, 65, ■   **8** 30, 40, 50, ■

9 65, 75, 85, ■   **10** 0, 10, 20, ■

11 Robin counts by 10. She writes 35, 45, 55. What is the next number?

12 Juan counts money. He says 5, 15, 25. What is the next number?

Answer Box ....

A	B	C	D	E	F
80	95	40	90	45	60

G	H	I	J	K	L
50	65	75	30	35	70

Objective: Continue a number pattern by skip counting by 10 to 100.

Where Is It?

Find the answer.

1 ☐ is above the .

2 ☐ is below the .

3 ☐ is below the .

4 ☐ is above the .

5 ☐ is below the .

6 ☐ is above the .

7 ☐ is below the .

8 ☐ is above the .

9 ☐ is above the .

10 ☐ is below the .

11 ☐ is above the .

12 ☐ is below the .

Answer Box

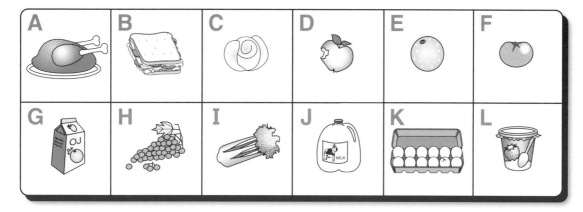

Objective: Identify the object that is above or below a given object.

What Time Is It?

Example

What time is it? The time is **3:30**.

What time is it?

1

2

3

4

5

6

7

8

9

10

11

12

Answer Box

A	B	C	D	E	F
8:30	3:30	7:30	5:30	11:30	12:30
G	**H**	**I**	**J**	**K**	**L**
9:30	1:30	10:30	4:30	6:30	2:30

Objective: Tell time to the half-hour, using an analog clock.

Who Is the Winner?

Soccer Games Won	
Team	Number of Games Won
Team A	10
Team B	12
Team C	8
Team D	9
Team E	11

How many games did the team win?

1 Team A

2 Team D

3 Team B

4 Team E

Who won more games?

5 Team C or Team D

6 Team E or Team A

7 Team D or Team A

8 Team B or Team E

How many more games did:

9 Team B win than Team A?

10 Team E win than Team C?

11 Team B win than Team C?

12 Team E win than Team A?

Answer Box

A	B	C	D	E	F
11	10	4	Team D	Team E	9
G	H	I	J	K	L
Team A	3	12	2	1	Team B

Objective: Read and interpret data given in a table.

Formal Tests Answer Key, Level 1

Counting 1–100, Book 1

1. 9
2. 8
3. 33
4. 15
5. 12
6. 42
7. Answers will vary.
   Sample answer: A b Ⓒ d
8. 46
9. Answers will vary.
   Sample answer: 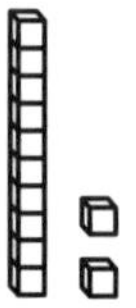
10. 35

Let's Estimate, Book 2

1. Hip
2. lizard
3. tiger
4. pool
5. cat
6. Mike
7. Otis
8. 10
9. 18
10. 35

Let's Add, Book 3

1. 9
2. 7
3. 7
4. nine (or 9)
5. **a.** 15
   b. 16
6. 2
7. 9
8. **a.** 7 and 3
   b. 14
9. **a.** 4+8 =
   b. 12
10. **a.** 1
    b. 5
    c. 4
    d. 7

Let's Subtract, Book 4

1. 4
2. 9
3. Answers will vary.
   Sample answer:
4. 4
5. 8
6. 4
7. four or 4
8. 9
9. 5
10. 9

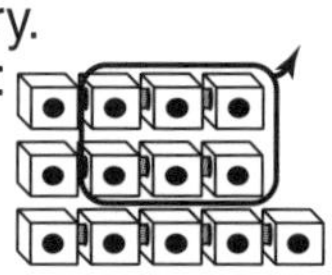

All About Patterns, Book 5

1.
2. 80
3. Answers will vary.
   Sample answer:

4. dog or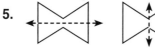
5. circle or ◯ 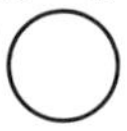
6. 35
7. –
8. +
9. 12
10. 4

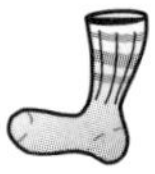

All About Shapes and Measures, Book 6

1. turkey
2. OJ or orange juice
3. 👟 
4. **a.** Sample drawing: △ 
   b. triangle
5. ⋈⟷ ⋈↕
6. 12

(Continued, right column)

7. 10 inches
8. 1 kilogram
9. fish tank
10. 20° F

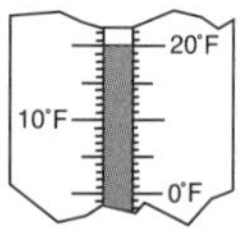

Time and Money, Book 7

1. Answers will vary. Sample
   answer: set the table for
   dinner and brush
   your teeth
2. 7:00
3. 4:30
4. August
5. January 4
6. 65¢
7. 78¢
8. 21¢
9. 28¢
10. Check students' work.
    Any amount from 21¢ and
    31¢ will suffice.

Tallies and Tables, Book 8

1. Answers will vary.
   Sample answer: flying things
2. Answers will vary.
   Sample answer: animals
3. Answers will vary.
   Sample answer: flying
   animals
4. 5
   ̶|̶|̶|̶|̶ ̶|̶|̶|̶|̶ |
5. 
6. 7
7. Mr. Chang
8. 4:15
9. Mine Street
10. Answers will vary. Sample
    answer: Change apple
    to girl.

Counting 1—100

Name: _____ Date: _____

1 How many in all? _____ 🧸🧸🧸🧸🧸 🧸🧸🧸🧸

2 Which number comes just after 7? _____

3 Which number comes just before 34? _____

4 Which number comes between 14 and 16? _____

5 Circle the greater number. 12 or 9

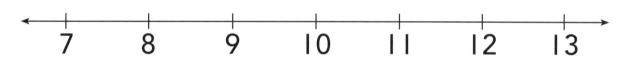

6 Circle the greater number. 38 or 42

7 Write 4 letters in a line. Circle the third letter. _____

8 How many? _____

9 Draw 12 blocks.

10 What number is 10 more than 25? _____

Let's Estimate

Name:_____ Date:_____

Circle the answer.

1 Who is tallest?

Kiko Tub Sag Hip Lob Doll Po

2 Which is longest? Frog Turtle Lizard

3 Which is heavier? Skunk Tiger

4 Which one holds more? Pail Pool

5 Which one is about **8** pounds? Cat Brush

Look at the chart.

6 Who has the most pennies?

_____

7 Who has **3** pennies?

_____

Our Pennies

Mike Ryan Otis Suki

Each 🪙 stands for 1 penny.

How many?

8 _____

9 _____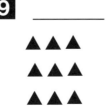

10 _____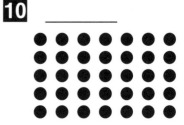

Let's Add

Name:_____ Date: _____

1 How many in all? _____

2 Find the missing number.

?	
3	4

3 plus 4 equals _____.

Find the sum.

3 $2 + 5 =$ _____

4 five plus four equals _____

5 **a.**
$$\begin{array}{r} 8 \\ + 7 \\ \hline \end{array}$$

b.
$$\begin{array}{r} 9 \\ + 7 \\ \hline \end{array}$$

Find the missing number.

6 $6 +$ _____ $= 8$

7 $8 +$ _____ $= 17$

8 $7 + 4 + 3$

a. Which two of these numbers make 10? _____

b. Find the sum. _____

9 **a.** Write an addition sentence to solve the problem.

4 .

8 more come. _____

b. How many in all? _____

10 Write the number that will make 10.

a. $9 +$ _____   **b.** $5 +$ _____   **c.** $6 +$ _____   **d.** $3 +$ _____

Let's Subtract

Name: _____ Date: _____

1 How many are left? _____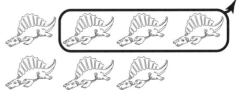

2 Han has 12 . He gives 3 away. How many are left?

_____

3 Draw a picture showing 13 – 6.

Find the missing number.

4

12	
?	8

12 minus _____ equals 8.

5 8 + 7 = 15

15 – 7 = _____

6 18 – 4 = 14

18 – 14 = _____

7 Eleven minus seven equals _____.

Find the difference.

8 16 – 7 = _____   **9**
$$\begin{array}{r} 18 \\ -13 \\ \hline \end{array}$$

10
$$\begin{array}{r} 17 \\ -\ 8 \\ \hline \end{array}$$

All About Patterns

Name: _____   Date: _____

What comes next?

1  _____

2 40, 50, 60, 70, _____

3 Make a repeating pattern. Use at least **2** different shapes.

Fill in the blank.

4

5 [cylinder] ◯ ◯ [cylinder] ◯ ◯ [cylinder] —— ◯ [cylinder]

6 5, 15, 25, _____, 45, 55

Fill in + or –.

7 9 § 2 = 7 _____   **8** 5 § 7 = 12

9 Use the first fact to find the missing sum.

6 + 5 = 11, so 6 + 6 = _____

10 Use the first fact to find the missing difference.

8 – 3 = 5, so 8 – 4 = _____

All About Shapes and Measures

Name:_____ Date: _____

Name the object.

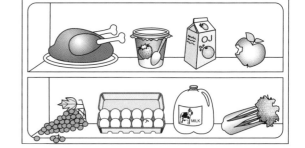

1 above the  _____

2 to the right of the _____

---

3 Circle the shape that does not belong.

4 **a.** Draw a shape with 3 sides.

   b. Name it. _____

5 Draw a line so the shape shows two equal parts.

6 How many 🐞 long is it? _____

To answer questions 7 and 8, circle the best estimate for the book.

7 10 inches or 5 feet

8 1 kilogram or 10 kilograms

---

9 Which one holds 20 liters?

10 It is snowing. Is it 20° F or 40° F?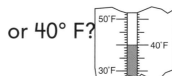

Time and Money

Name:_____ Date: _____

1 Name 2 things that take less than 15 minutes to do.

_____   _____

What time is it?

2 _____    **3** _____

4 What month comes after July? _____

5 What is the date for Thursday? _____

JANUARY						
Sunday	Monday	Tuesday	Wednesday	Thursday	Friday	Saturday
	1	2	3	4	5	6

How much money?

6  _____

7  _____

8 Two nickels, a dime, and a penny? _____

9 Find the cost of 2 pens and a chocolate bar.

 11¢   6¢

10 Write an amount between 20¢ and 32¢. _____

Draw coins to show that amount.

Tallies and Tables

Name:_____ Date: _____

Use the diagram to answer questions 1–4.

1 Name the group in the left circle. _____

2 Name the group in the right circle. _____

3 Name the group in both circles. _____

4 How many are in the right circle? _____

Use the pictograph to answer questions 5–7.

Number of Fish	
Ms. Hastings	🐟🐟🐟🐟🐟🐟🐟
Mr. Chang	🐟🐟🐟🐟🐟🐟🐟🐟🐟🐟🐟🐟🐟
Mr. Costa	🐟🐟🐟🐟
Each 🐟 stands for 1 fish.	

5 Use tally marks to show how many fish Mr. Chang has. _____

6 How many fish does Ms. Hastings have? _____

7 Who probably has the biggest fish tank? _____

Use the schedule to answer questions 8 and 9.

8 What time does the bus stop at River Road? _____

9 Where is the bus at 4:12? _____

10 Change one word to make the sentence real. The apple kicked the ball. _____

Bus Stops			
Street	Time	Street	Time
Broadway	4:05	River Road	4:15
Main Street	4:08	Oak Street	4:18
Mine Street	4:12	Maple Street	4:20

© ETA/Cuisenaire®

Correlation of Tests to Student Activity Books

Test Questions	Topic	Book Page
Counting 1–100, Level 1, Book 1		
1	Counting Whole Numbers: numbers to 13	1–4, 10, 11
2, 5	Comparing and Ordering Whole Numbers: numbers to 12	5–11
3, 4, 6, 10	Comparing and Ordering Whole Numbers: numbers to 100.	5–11, 14–24, 27–32
7	Counting Whole Numbers: ordinal numbers	12, 13
8–9	Naming Whole Numbers: place-value models to 100	14–20, 25, 26
Let's Estimate, Level 1, Book 2		
1, 3	Estimating Measurements: length	1–6, 14, 15
2, 5	Estimating Measurements: weight	7–10, 13
4	Estimating Measurements: capacity	11, 12
6	Using Models of Addition and Subtraction: groups of objects	17, 18, 20, 21, 23–32
7	Using Models of Addition and Subtraction: pictographs	19, 22
8-10	Skip Counting to Find Sums: skip count by 2, 3, 4, 5	23–32
Let's Add, Level 1, Book 3		
1	Using Models of Addition and Subtraction: counting totals	1, 2
2	Part-Part-Total Models: sums, addends, and addition sentences	3–6
3	Addition Facts to 18: adding 0–5 and addition expressions	9–15
4	Addition Facts to 18: using words	25
5	Addition Facts to 18: vertical form and addition facts	18, 19, 28–30
6, 7	Addition Facts to 18: addition sentences, missing addends and sums	3-8, 16, 17, 20, 21, 22
8	Addition Facts to 18: using making a ten and three addends	23, 24, 31–32
9	Addition Facts to 18: addition sentences	16, 17, 20
10	Addition Facts to 18: using making a ten	24
Let's Subtract, Level 1, Book 4		
1	Using Models of Addition and Subtraction: counting the number being taken away and the number left	1–3
2, 3	Subtraction Facts to 18: subtraction expressions	10–17, 26, 27
4	Part-Part-Total Models: missing differences and subtrahends	5, 6, 8, 20, 21
5	Subtraction Facts to 18: using addition facts	4, 18, 19, 23, 24
6	Subtraction of Whole Numbers: using related facts	20
7	Subtraction Facts to 18: using words	25
8	Subtraction Facts to 18: subtraction sentences	7, 9, 22
9, 10	Subtraction Facts to 18: vertical form	28–32
All About Patterns, Level 1, Book 5		
1	Shape Patterns: real-life objects	1, 4
2	Skip Counting Patterns: skip counting to 100	11–16
3	Shape Patterns: 2- and 3-dimensional shapes	6–8
4	Identifying Shapes: similar objects	2–3
5	Identifying Shapes: similar geometric shapes	5
6	Number Patterns: counting forward and backward/repeating patterns	9, 10, 21, 22
7, 8	Function Machines: rules	19, 20
9	Addition and Subtraction Patterns: addition facts	23–28
10	Addition and Subtraction Patterns: subtraction facts	29–32

Correlation of Tests to Student Activity Books

Test Questions	Topic	Book Page
All About Shapes and Measures, Level 1, Book 6		
1	Spatial Relationships: above or below	1, 2
2	Spatial Relationships: left or right	3, 4
3	Congruencies and Similarities: matching groups of objects	5
4	Identifying 2- and 3-dimensional Shapes: descriptions of 2- and 3-dimensional shapes	8–15
5	Congruencies and Similarities: congruent 2-dimensional figures	6, 7, 16, 17
6	Measuring Length: nonstandard units	18, 19
7	Estimating Length, Weight, Capacity, and Mass: length in customary units	22, 26, 27
8	Estimating Length, Weight, Capacity, and Mass: mass	24, 26–28
9	Estimating Length, Weight, Capacity, and Mass: capacity in metric units	25, 29
10	Temperature: estimating	31, 32
Time and Money, Level 1, Book 7		
1	Time: more/less time	1, 2
2	Time: telling time to the hour	5–7, 10–12
3	Time: telling time to the half-hour	3, 4, 7–9, 12
4	Calendar: months	14, 15
5	Calendar: days and dates	13, 14
6-8	Money: counting coins	16–25, 26, 30, 31
9	Money: buying items	27, 30, 31
10	Money: comparing and ordering	28, 29, 32
Tallies and Tables, Level 1, Book 8		
1, 2	Sorting Information: objects by category	3–6
3, 4	Sorting Information: Venn diagrams	8
5	Reading and Interpreting tables: tally marks	9, 10
6	Reading and interpreting Graphs: pictographs and single bar graphs	16–25
7	Certain, Possible, and Impossible Events: likely and unlikely events	28, 29
8, 9	Reading and Interpreting Tables: schedules	14, 15
10	Certain, Possible, and Impossible Events: real and not real events	30, 31

VersaTiles® Student Record Chart

Student Name: _____

Book Number: _____

Book Title: _____

Page	Date Completed	Page	Date Completed	Page	Date Completed
1		9		17		25	
2		10		18		26	
3		11		19		27	
4		12		20		28	
5		13		21		29	
6		14		22		30	
7		15		23		31	
8		16		24		32	

VersaTiles® Student Record Chart

Student Name: _____

Book Number: _____

Book Title: _____

Page	Date Completed	Page	Date Completed	Page	Date Completed
1		9		17		25	
2		10		18		26	
3		11		19		27	
4		12		20		28	
5		13		21		29	
6		14		22		30	
7		15		23		31	
8		16		24		32	

VersaTiles® Class Record Chart

Book Number: _____

Book Title: _____

Student	PAGES COMPLETED																															
	1	2	3	4	5	6	7	8	9	10	11	12	13	14	15	16	17	18	19	20	21	22	23	24	25	26	27	28	29	30	31	32

VersaTiles® Student Self-Assessment Sheet

Student Name: _____ Date: _____

Book Number: _____ Book Title: _____

Answer each question below. Think about how you felt about your work in this VersaTiles book. There are no wrong answers. Your answers will not be graded.

1. What did you like most about this book?

_____

_____

_____

2. What did you like least about this book?

_____

_____

_____

3. What did you learn from this book?

_____

_____

_____

4. Is there any part of the book you wish you understood better?

_____

_____

_____

Certificate of Completion

GREAT JOB!

This certificate is presented to

upon successful completion of

VersaTiles® Book

Teacher's Signature

VersaTiles® Work Slate

© ETA/Cuisenaire®

Student Name: _____

1	**2**
3	**4**
5	**6**
7	**8**
9	**10**
11	**12**

VersaTiles® Template

Activity Title: _____   **Student Name:** _____

Directions: _____

1

2

3

4

5

6

7

8

9

10

11

12

Answer Box ·

A	B	C	D	E	F
G	H	I	J	K	L

90   Objective: _____

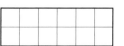

VersaTiles® Template

Activity Title: _____   Student Name: _____

Directions: _____

1

2

3

4

5

6

7

8

9

10

11

12

Answer Box ·

A	B	C	D	E	F
G	H	I	J	K	L

Objective: _____   91

Patterns and Pattern Codes for Creating VersaTiles Activities

Pattern 1

1	L
2	H
3	I
4	K
5	C
6	G
7	E
8	B
9	J
10	F
11	D
12	A

Pattern 2

1	J
2	H
3	E
4	C
5	K
6	A
7	I
8	B
9	L
10	D
11	F
12	G

Pattern 3

1	B
2	F
3	I
4	G
5	C
6	K
7	A
8	L
9	D
10	H
11	J
12	E

Pattern 4

1	H
2	J
3	K
4	G
5	F
6	I
7	B
8	C
9	L
10	A
11	E
12	D

Pattern 5

1	D
2	F
3	A
4	C
5	G
6	E
7	I
8	L
9	B
10	J
11	H
12	K

Pattern 6

1	E
2	C
3	B
4	F
5	H
6	D
7	L
8	I
9	A
10	K
11	G
12	J

Pattern 7

1	K
2	G
3	J
4	L
5	D
6	H
7	F
8	A
9	I
10	E
11	C
12	B

Pattern 8

1	D
2	B
3	E
4	C
5	L
6	A
7	J
8	G
9	F
10	I
11	K
12	H

Pattern 9

1	G
2	K
3	J
4	H
5	C
6	L
7	A
8	F
9	I
10	B
11	D
12	E

Pattern 10

1	A
2	C
3	F
4	B
5	K
6	D
7	G
8	J
9	E
10	H
11	L
12	I

Pattern 11

1	J
2	L
3	G
4	I
5	A
6	K
7	D
8	F
9	H
10	C
11	B
12	E

Pattern 12

1	L
2	D
3	G
4	K
5	B
6	C
7	F
8	J
9	H
10	E
11	A
12	I

Patterns and Pattern Codes for Creating VersaTiles Activities

Pattern 13

#	Code
1	F
2	B
3	D
4	E
5	J
6	A
7	K
8	H
9	C
10	L
11	I
12	G

Pattern 14

#	Code
1	J
2	A
3	F
4	I
5	L
6	B
7	C
8	G
9	E
10	D
11	K
12	H

Pattern 15

#	Code
1	G
2	K
3	J
4	H
5	C
6	L
7	B
8	E
9	I
10	A
11	D
12	F

Pattern 16

#	Code
1	B
2	J
3	E
4	A
5	L
6	I
7	H
8	C
9	F
10	G
11	K
12	D

Pattern 17

#	Code
1	I
2	K
3	H
4	J
5	A
6	L
7	D
8	F
9	G
10	C
11	B
12	E

Pattern 18

#	Code
1	E
2	D
3	B
4	F
5	G
6	C
7	K
8	J
9	A
10	L
11	H
12	I

Pattern 19

#	Code
1	F
2	H
3	J
4	K
5	C
6	A
7	E
8	B
9	I
10	L
11	D
12	G

Pattern 20

#	Code
1	J
2	B
3	E
4	I
5	K
6	G
7	D
8	H
9	L
10	C
11	F
12	A

Pattern 21

#	Code
1	G
2	E
3	I
4	B
5	C
6	L
7	H
8	K
9	J
10	A
11	D
12	F

Pattern 22

#	Code
1	A
2	D
3	F
4	H
5	L
6	C
7	B
8	J
9	K
10	G
11	E
12	I

Pattern 23

#	Code
1	C
2	K
3	G
4	D
5	B
6	L
7	J
8	F
9	H
10	I
11	A
12	E

Pattern 24

#	Code
1	L
2	C
3	H
4	K
5	A
6	D
7	F
8	J
9	G
10	E
11	B
12	I

Notes

Notes

Notes